Genes, Girls, and Gamow

Jim Watson in Moscow at the International Biochemical Congress, 1961.

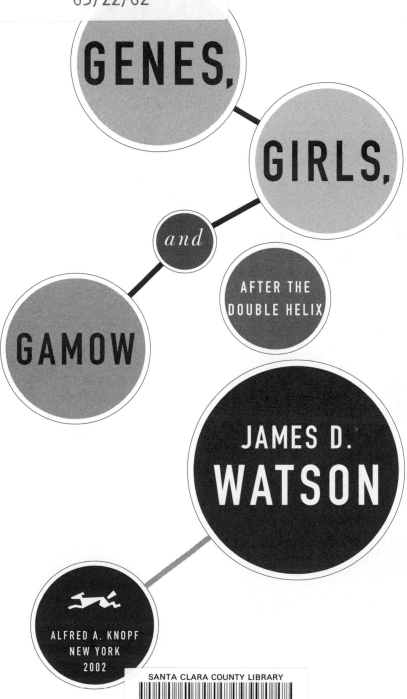

GENES,

GIRLS,

and

GAMOW

AFTER THE
DOUBLE HELIX

JAMES D.
WATSON

ALFRED A. KNOPF
NEW YORK
2002

THIS IS A BORZOI BOOK
PUBLISHED BY ALFRED A. KNOPF

Originally published in Great Britain in slightly different
form by Oxford University Press, London, 2001.

Knopf, Borzoi Books, and the colophon are
registered trademarks of Random House, Inc.

Library of Congress Cataloging-in-Publication Data

Watson, James D., 1928–
Genes, girls, and Gamow: after The Double Helix/James D. Watson.—1st ed.
p.   cm.
"A Borzoi book."
ISBN 0-375-41283-2 (alk. paper)
1. Watson, James D., 1928– 2. Molecular biologists—United States—Biography.
I. Watson, James D., 1928– Double Helix.   II. Title.

QH506 .W399 2002
572.8'092—dc21   2001038543
[B]

Manufactured in the United States of America
First American Edition

To Celia Gilbert

It is a truth universally acknowledged, that a single man in possession of a good fortune, must be in want of a wife.

—Jane Austen, *Pride and Prejudice*

Foreword

It is reasonable for one to observe the universe about one and report what one sees. After all, there are countless reporters and correspondents who do just that for the media. Also, the first step in the scientific method is to observe and report one's observations.

Psychological interpretations, motivations, and intents are rather more dangerous.

As the chemist Harry Kroto clearly pointed out to me a year ago, "Everything is subjective." In this book the subject is Jim Watson. There are many other players, real people, a good many of whom will be unhappy with the book (the Victims). Without them, however, there would be no book.

The information available to me indicates that the stimulus for this book is girls, particularly Christa Mayr. Being slow and egocentric, I did not realize that romantic Jim had such great problems about girls, although there was much evidence available. My problems were rather the opposite.

As a work of reference to what actually happened, this book is unreliable. There are many mistakes and errors of fact. Some of these are minor and refer to things that Jim did not observe directly.

I suggest that every coffee table and hairdresser have this book, the latter for the ladies to read under the hairdryer what is in many ways an entertaining book.

As unappointed leader of the Victims, I hope they will forgive or at least be lenient with both me and Jim.

PETER PAULING

Preface

The chase for the double-helical structure of DNA was an adventure story in the best sense. First, there was a pot of scientific gold to be found—possibly very soon. Second, among the explorers who raced to find it, there was much bravado, unexpected lapses of reason, and painful acceptances of the fates not going well. The early 1950s were not times to be cautious but rather to run fast whenever a path opened up—nuggets of gold might be lying exposed over the next hill. As one of the winners with a fortune much, much bigger than I ever dared hope for, I could not stop moving. There was more genetic loot to be located, and not joining in the further hunt would make me feel old. Out there was the genetic code—the "Rosetta Stone of Life"—that would tell us the rules by which genetic information encoded within DNA molecules is translated into the language of proteins, the molecular workhorses of all living cells.

From the start, the best path towards the genetic code seemed in or near a still-mysterious molecule called ribonucleic acid, or RNA. Although quite distinct from DNA, it was built along the same lines and might also encode genetic information. In the spring of 1953, I had no idea what RNA looked like, and this book, in part, is the story of its pursuit. Conceivably mere inspection of RNA's three-dimensional form would tell us the genetic code and set us towards the molecular machinery that uses its rules to translate the language of DNA into the language of proteins.

In this search I again often had Francis Crick with me. But fates sometimes placed us thousands of miles apart, and many of my steps towards unraveling RNA were in the company of newer friends. For the most part, they had seen forests of seemingly impenetrable molecules before and knew approximately what clothes to wear and the tools needed to cut the thickets ahead. Quite different was the truly bizarre explorer, the Russian-born, George Gamow. Theoretical physicist *extraordinaire,* and a six-foot, six-inch giant to boot, "Geo," as he ended his letters, defied conventional description with his penchant for tricks that masked a mind that always thought big. Together we were to form a club with a tie that he designed and called the RNA Tie Club. To its 20 members, Francis Crick sent his famous 1955 "Adaptor Hypothesis" that he never published elsewhere. Our club became part of the history of molecular biology.

For years I have wanted to write about how the RNA Tie Club came into existence, inserting Geo's oft-illustrated, wacky letters into the intellectual climate that surrounded the spiritual upheaval among biologists after the discovery of the double helix. I could have restricted the story to scientific issues, but have placed it instead within the context of my own personal life, itself strongly influenced by the lives of my friends. The story starts when I was an unmarried 25-year-old and thought more about girls than genes. It is as much a tale of love as of ideas.

Like with *The Double Helix,* I try to capture the spirit of my youth and purposely do not make reflective judgment on where I was going right or wrong. Recapitulating the essence of times long ago, however, risks repeating long-held mismemories. Luckily and graciously put at my disposal were some 60 letters that I wrote to Christa Mayr Menzel between July 1953 and December 1955. In rereading them, I found an almost diary-like description of the people then entering my life as well as my scientific brainstorms of the moment. I also faithfully kept all the letters that I received from other close acquaintances of that period.

By purposely not judging my actions of those past days, I risk upsetting readers who want to see me as I am today, rather than as the inexperienced and more self-centered person I once was. There will be other readers, possibly not so harsh on my character either then or now, who nonetheless feel that many of the personal facts I write below are not worth being passed on to the future. Almost everyone goes awry in some aspect of their human lives, and what I describe may not be that unusual. But for better or worse, I and my friends were present at the birth of the DNA paradigm—by any standard one of the great moments in the history of science, if not of the human species. In this way we were unique players in a momentous drama.

Thus there will be many readers wanting to know better what actually happened in our lives. This story is no second romp to a double-helix-like discovery, but it has Geo Gamow to keep us on our toes. A giant imp, jumping from atoms to genes to space travel, he was simultaneously there to admire when clever and to comfort when his life was going backwards. Perhaps wisely, Geo never counted on ever finishing the big chases he started. So he always sought fun on the way. Thinking now back on my life, Geo was a much wiser individual than I first judged him.

Acknowledgments

The turning of my first draft into this more taut version benefited from comments from many friends. All the major participants were sent one or more drafts to let them tell me where their memories of events differed from mine. Donald Caspar, Francis Crick, Paul Doty, Celia Gilbert, Hugh Huxley, Leslie Orgel, Peter and Alicia Pauling, and Baroness Janet Stewart Whitaker all so helped me. In addition, I benefited from comments by John Cairns, Richard Dawkins, Nancy Hopkins, Gordon Lish, Cynthia MacKay, Victor McElheny, Benno Müeller-Hill, Mark Ptashne, Matt and Anya Ridley, Peter Sherwood, Jan Witkowski, and Norton Zinder.

As I still write by hand, without my very competent assistant Maureen Berejka's help in deciphering my writing and transferring it to her word processor, this book could not have appeared. I must also thank Ramah McKay for correcting my grammar and and urging me to shorten unnecessarily wordy sentences.

In Michael Rodgers of Oxford University Press and George Andreou of Knopf, I have benefited from seasoned editors of taste. They not only know words best unsaid, but equally important, paragraphs best left in to faithfully record how the post double-helix world unfolded.

Lastly, I must acknowledge my wife Liz's continued warm support and encouragement going back to when she typed early chapters during summers at our then house on Vincent Square in London. Many subsequent chapters were written in Oxford in the spring of 1994 when I was the Newton Abraham Professor attached to Lincoln College. I wish to thank its Rector and Fellows for making our stay there so pleasant. Not wanting then to lose Oxford from our lives, Liz and I later bought a ground floor flat with garden on Northmoor Road. Its solitude let me write the concluding chapters.

J. D. WATSON
November 2001
Cold Spring Harbor, N.Y.

Cast of Characters

Beadle, George ("**Beets**") Geneticist (b. 1903), Head of the Biology Division at Caltech from 1946 to 1961. In 1958, he shared the Nobel Prize with Edward Tatum for experiments at Stanford University connecting genes and proteins.

Benzer, Seymour Brooklyn-born physicist-turned-geneticist (b. 1921); on the faculty at Purdue University between 1945 and 1967 before moving to Caltech as Professor of Biology.

Bernal, John Desmond Irish crystallographer (b. 1901) who, in the 1930s, supervised the Cambridge Ph.D. theses of Max Perutz and Dorothy Crowfoot Hodgkin; known for his left-wing politics—his office at Birkbeck College London had a peace dove drawn on its walls by Picasso.

Bohr, Niels Danish physicist (b. 1885) who was the first to postulate that electrons circle the nucleus in fixed quantum orbits, for which he received the Nobel Prize for Physics in 1922. In the 1920s and 30s, his Copenhagen Institute attracted the world's cleverest theoreticians—its temporary inhabitants included George Gamow, Lev Landau, and Werner Heisenberg.

Bragg, Sir (William) Lawrence Australian-born (1890) British physicist. His father, Sir William Henry, and he were the effective founders of X-ray crystallography through their 1912 law on how crystals diffract X-rays. For Bragg's law, they received the 1915 Nobel Prize for Physics. In 1938, W. L. Bragg succeeded Ernest Rutherford as the head of the Cavendish Laboratory in Cambridge.

Brenner, Sydney Raised in South Africa (b. 1927), he obtained a medical qualification from Witwatersrand University in Johannesburg before moving, in 1951, to Oxford to do research for his Ph.D. under the supervision of the chemist Cyril Hinshelwood. In 1958, he moved to Cambridge to work with Francis Crick.

Bullard, Belinda Cambridge undergraduate (b. 1936) at Girton College reading biochemistry; the daughter of the noted geophysicist Sir Edward Bullard, the scion of a rich Norwich brewing family.

Calvin, Melvin University of California chemist (b. 1911) whose unraveling of key steps in the assimilation of carbon dioxide during photosynthesis led to his 1961 Nobel Prize for Chemistry.

Caspar, Don American biophysicist (b. 1930); after X-ray studies on tobacco mosaic virus for a Ph.D. at Yale, he was a postdoctoral fellow at Caltech during 1955–6. The next year he went to the University of Cambridge to study spherical plant viruses.

Chargaff, Erwin Czech-American biochemist (b. 1905) who studied in Vienna, in the U.S. at Yale, and also in Berlin. He returned to the U.S. and worked from 1935 at the College of Physicians and Surgeons of Columbia University. In the early 1950s, he found that the DNA bases adenine and thymine are present in equal numbers as are the bases guanine and cytosine.

Crick, Francis Born near Northampton in 1916. After training as a physicist at University College London, he did scientific research for the British Admiralty during World War II. In 1947, he moved to Cambridge to learn biology and two years later joined the Medical Research Council Unit for the Study of the Molecular Structure of Biological Systems at the Cavendish Laboratory, where he became a Ph.D. student of Max Perutz. The unit later grew into the MRC Laboratory of Molecular Biology (upon moving to a new site at Addenbrooke's Hospital in 1962).

Cullis, Ann Superb technician (b. 1931) for Max Perutz at the Cavendish Laboratory in Cambridge during the mid-1950s.

Delbrück, Max German theoretical physicist (b. 1906) who was associated for several years with Niels Bohr in Copenhagen before returning to Berlin to work with Lisa Meitner at the Kaiser Wilhelm Institute of Chemistry. He then became interested in genetics and moved in 1937 to Caltech, where he began working with bacterial viruses (phages). In 1941, he married Mary ("Manny") Bruce, whose father was an engineer with Cyprus Mines. In 1969, he shared the Nobel Prize for Physiology or Medicine with Alfred Hershey and Salvador Luria.

Demerec, Milislav Yugoslav geneticist (b. 1895) who studied for a Ph.D. in plant breeding at Cornell University. He moved to the Cold Spring Harbor Laboratory in 1923 and became its Director in 1941.

Donohue, Jerry A theoretical chemist (b. 1920) at Caltech with Linus Pauling, he and his wife Pat were at the Cavendish Laboratory in Cambridge in 1952–3. There he occupied an office with Francis Crick, JDW, and Peter Pauling, giving the opinion in late February 1953 that guanine and thymine would have keto, not enol, conformations.

Doty, Paul American physical chemist (b. 1920) whose research turned to DNA after he joined the Harvard Chemistry Department in 1948. In 1954, he married his Ph.D. student, Helga Boedtker.

Dulbecco, Renato Qualified as a doctor in Italy (b. 1914); emigrated to the U.S. in 1947 to work with Salvador Luria at Indiana University on bacterial viruses. He moved to Caltech in 1949 to extend phage methodologies to animal viruses. He shared the 1975 Nobel Prize for Physiology or Medicine for his work on DNA tumor viruses.

Dunitz, Jack Scottish-born (1923) and -trained X-ray crystallographer who moved to Oxford as a postdoctoral fellow with Dorothy Crowfoot Hodgkin in 1946. Afterwards he spent several years at Caltech where he met Alex Rich, later joining him at the National Institutes of Health (NIH). In 1957, he was appointed Professor of Chemical Crystallography in Zurich.

Ephrussi, Boris Russian-born (1901), French-educated geneticist who was in the U.S. during World War II. After the war he returned to Paris as Professor of Genetics at the Sorbonne, focusing on the genetics of yeast. In 1949, he married his second wife, the American microbiologist Harriet Taylor, whom he met at the 1946 Cold Spring Harbor Symposium.

Feynman, Richard ("Dick") American physicist (b. 1918); after getting his Ph.D. from Princeton, he taught physics at Cornell before becoming, in 1952, Professor of Theoretical Physics at Caltech. A founder of quantum electrodynamics, he shared the 1965 Nobel Prize for Physics.

Fraenkel-Conrat, Heinz German-born (1910), he studied medicine in Breslau and did his Ph.D. in chemistry at the University of Edinburgh. In 1952, he joined Wendell Stanley's Virus Laboratory at the University of California, Berkeley, where he worked on plant viruses.

Franklin, Rosalind Cambridge-educated physical chemist (b. 1921). After four years in Paris (1947–51), she joined the Medical Research Council Biophysics Unit at King's College London. There she located the phosphate atoms of DNA on the outside surface and discovered DNA's "B" form. In 1953, she transferred to J. D. Bernal's lab at Birkbeck College London.

Gamow, George ("Geo") Russian-born theoretical physicist (born in Odessa, 1904). He obtained his Ph.D. in 1928 from St. Petersburg University and then spent the next three years in Copenhagen and Cambridge before returning to Russia in 1931. A meeting in Belgium gave him and his wife (Rho) the opportunity to move to Washington, D.C., where, between 1934 and 1956, he was Professor of Physics at George Washington University.

Gierer, Alfred German biochemist (b. 1929). He trained to work on tobacco mosaic virus in the Max Planck Institute for Virus Research in Tübingen, where he was a protégé of Gerhard Schramm.

Gilbert, Celia Smith College–educated daughter of the journalist I. F. Stone and wife of Walter Gilbert.

Gilbert, Walter ("Wally") Boston-born (1932); after graduating from Harvard, he went to the University of Cambridge to study for his Ph.D. in theoretical physics supervised by Abdus Salem (who won the 1979 Nobel Prize for Physics). Returning to Harvard in 1956, he taught theoretical physics until 1964, when he became Associate Professor of Biophysics. In the early 1970s, he independently developed a powerful way for sequencing DNA that led to his sharing the 1980 Nobel Prize for Chemistry with Fred Sanger and Paul Berg.

Griffiths, Sheila Raised in Wales (b. 1928), daughter of James Griffiths, a Labour Party MP and member of the Labour Government 1945–51. In the summer of 1952 she met JDW at a village in the Italian Engadine. She returned to England in the spring of 1953 and in the following year married a young historian, Roy Pryce, whom she had met in Rome.

Haldane, J. B. S. Geneticist (b. Oxford 1892), educated at Eton and New College, Oxford, brother of the writer Naomi Mitchison. He displayed brilliance and eccentricity both as an experimentalist and theoretical geneticist, first in Cambridge and later at University College London. Long a member of the British Communist Party, he emigrated in 1956 to India where he died in 1964.

Hershey, Alfred American chemist (b. 1908) who initiated studies on phage at Washington University. In 1950, he moved to the Cold Spring Harbor Laboratory where, with Martha Chase, he showed that the DNA component of phages was its genetic component. For this work, he received a Nobel Prize in 1969 (shared with Max Delbrück and Salvador Luria).

Hinshelwood, Sir Cyril English physical chemist (b. 1897), educated at Balliol College, Oxford; was Oxford's Dr. Lee's Professor of Chemistry between 1937 and 1964. He became an authority on chemical kinetics for which he shared the 1956 Nobel Prize for Chemistry with Nikolai Semenov. His book *The Chemical*

Kinetics of the Bacterial Cell (1946) generated widespread disapprobation from geneticists for its Lamarckian outlook.

Hodgkin, Dorothy (née Crowfoot) British crystallographer born in Cairo (1910). She studied chemistry at Somerville College, Oxford and, later, as a Ph.D. student of J. D. Bernal in Cambridge, she was the first person to obtain X-ray diffraction patterns from protein crystals. In 1936, she moved back to Oxford where she used X-ray diffraction methods to establish the structure of penicillin. This work, together with her later elucidation of the structure of vitamin B_{12}, led to her Nobel Prize for Chemistry in 1964.

Huxley, Hugh English biophysicist (b. 1924) who began his studies on muscle contraction in Cambridge, at the Cavendish Laboratory, as a Ph.D. student of John Kendrew. From 1952 to 1954 he was a Commonwealth Fellow at MIT, where he began electron microscope studies that led to the sliding filament model of muscle contraction. After briefly coming back to the Cavendish Laboratory, he joined the Biophysics Department at King's College London in 1956. In 1961, he returned to Cambridge.

Huxley, Julian The grandson (b. 1887) of Darwin-proponent Thomas Henry Huxley and the brother of the novelist and essayist Aldous Huxley. After Eton and Balliol College, Oxford, he taught for three years at Rice University in Houston but returned to England during World War I. A prolific writer of books, he became the first Director General of UNESCO in 1946.

Jacob, François Medically trained French biochemist (b. 1920) whose later research on gene regulation at the Institut Pasteur in Paris with André Lwoff and Jacques Monod led to their joint Nobel Prize for Medicine or Physiology in 1965.

Kendrew, John Oxford-born (1917) molecular biologist. After Trinity College, Cambridge, he spent World War II in aviation operational research where he first met J. D. Bernal. He and Max Perutz were the first members of the Medical Research Council Unit for the Study of the Molecular Structure of Biological Systems at the Cavendish Laboratory in Cambridge, where he initiated his studies on the oxygen-carrying protein myoglobin.

Khorana, Gobind Born in India (1922), he was educated at Lahore and Liverpool universities. His studies on nucleic acid began when he was a postdoctoral fellow in Alexander Todd's Cambridge organic chemistry lab between 1948 and 1952. His studies on RNA at the Institute of Enzyme Research of the University of Wisconsin culminated in his enzymatic synthesis of specific RNA sequences that proved crucial to the establishment of the genetic code. For this work, he shared the 1968 Nobel Prize for Physiology or Medicine (with Marshall Nirenberg and Robert Holley).

Klug, Aaron Born in Lithuania (1926) and educated in South Africa, he obtained his Ph.D. in physics at the Cavendish Laboratory in Cambridge on studies of the structure of steel. In 1954, he joined Rosalind Franklin in London to work with her on the structure of tobacco mosaic virus. After her death in 1958, he continued to work at Birkbeck College until 1962 when he moved to the Medical Research Council Laboratory of Molecular Biology in Cambridge. For work done to establish crystallographic electron microscopy, he received the 1982 Nobel Prize for Chemistry.

Landau, Lev Russian theoretical physicist born in Baku, Azerbaijan (1908), and friend of George Gamow—both were in Copenhagen with Niels Bohr in 1930 when quantum mechanics was coming into existence. For his theories on liquid helium, he was awarded the Nobel Prize for Physics in 1962.

Ledley, Robert American mathematician (b. 1926) briefly interested in the logic of the genetic code.

Lewis, Elizabeth Vickery Born (1948) in Providence, Rhode Island; Radcliffe sophomore doing administrative work in JDW's lab in 1967.

Lewis, Julia Undergraduate (b. 1936) at Girton College, Cambridge in the mid-1950s, studying languages.

Luria, Salvador Medical graduate of Turin University (b. 1912). He studied phages at the Radium Institute in Paris in 1938, but when Italy entered World War II he fled to the U.S., where he continued his phage experiments at Columbia University's College of Physicians and Surgeons in New York. In 1943, he became a member of the faculty of Indiana University; in 1950 he moved to the University of Illinois and from there to MIT in 1958. He shared the 1969 Nobel Prize for Physiology or Medicine with Max Delbrück and Alfred Hershey.

Lwoff, André French microbiologist (b. 1902) at the Institut Pasteur who shared the 1968 Nobel Prize for Physiology or Medicine with François Jacob and Jacques Monod; a precise and playfully elegant speaker and writer.

Markham, Roy English biochemist (b. 1916), who innovatively focused on nucleic acids at his plant virology lab at the Molteno Institute in Cambridge.

Mayr, Christa Elder daughter (b. 1936) of Ernst and Gretel Mayr.

Mayr, Ernst German-born (1904) American ornithologist who presided over the Rothschild bird collection of the American Museum of Natural History in New York from 1933 to 1953. He and his wife Gretel were summer residents at the Cold Spring Harbor Laboratory from 1943 until 1953, when he was given a professorship at Harvard's Museum of Comparative Zoology.

Mayr, Susie Younger daughter (b. 1937) of Ernst and Gretel Mayr.

McMichael, Ann A blond American girl whose physician husband was learning molecular biology in Geneva in the summer of 1955.

Meselson, Matthew ("Matt") Born in Denver (1930), he was a Ph.D. student of Linus Pauling at Caltech between 1953 and 1956; later as a postdoctoral fellow, he and Franklin Stahl used ultracentrifugation to demonstrate that the two chains of DNA separate during its replication. In 1961, he moved to Harvard as Professor of Biology.

Metropolis, Nicholas Los Alamos computer whiz (b. 1915) who collaborated with George Gamow in 1954 to examine the randomness of amino acid sequences in proteins.

Mitchison, Avrion ("Av") Oxford-educated immunologist (b. 1928), son of Naomi and Dick Mitchison. As a Commonwealth Fellow, he researched in the U.S. between 1952 and 1954 at Indiana University and the Jackson Laboratory at Bar Harbor; afterwards he lectured in zoology at the University of Edinburgh.

Mitchison, Murdoch Cambridge-educated zoologist (b. 1921), who moved to the University of Edinburgh in 1952; older brother of Avrion. Married in 1947 to Rosalind Wrong, an Oxford historian.

Mitchison, Naomi ("Nou") Daughter (b. 1897) of famed Oxford physiologist John Scott Haldane and sister of J. B. S. Haldane. In 1916, she married her brother's close friend G. R. (Dick) Mitchison (b. 1892), barrister and Labour MP.

Monod, Jacques French microbial geneticist (b. 1910), who also excelled in music, sailing, and rock climbing. A member of the French Resistance during World War II, he became attached to the Institut Pasteur in 1945. There his charisma and intelligence quickly attracted an abundance of clever coworkers and sabbatical visitors. With André Lwoff and François Jacob, he was a recipient of the 1968 Nobel Prize for Physiology or Medicine.

Mulliken, Robert University of Chicago chemical physicist (b. 1897) whose molecular orbital approach to chemical bonding was never accepted by Linus Pauling. He was awarded the Nobel Prize for Chemistry in 1966.

Nirenberg, Marshall American biochemist (b. 1927) who at the National Institutes of Health in Bethesda discovered that synthetic polynucleotides promote the synthesis of sequence specific polypeptides. For this work he shared the 1968 Nobel Prize for Physiology or Medicine (with Gobind Khorana and Robert Holley).

Ochoa, Severo Spanish-born (1905) American biochemist in whose New York University laboratory Marianne Grunberg-Manago discovered the enzyme polynucleotide phosphorylase, which was later used to make synthetic RNA molecules. He was a joint winner with Arthur Kornberg of the 1959 Nobel Prize for Physiology or Medicine.

Oppenheimer, J. Robert American theoretical physicist (b. 1904) who oversaw in Los Alamos the construction of the first atomic bomb. In 1940, he married Katherine (Kitty) Harrison, previously married to the Caltech-associated, British-born physician Stuart Harrison.

Orgel, Leslie English theoretical chemist (b. 1927) and friend of Av Mitchison when they were both prize fellows at Magdalen College, Oxford; afterwards at Caltech he joined Linus Pauling's assembly of young theoreticians. In 1950, he married H. Alice Levinson who read medicine at Oxford.

Pauling, Linda Lively blond daughter (b. 1932) of Linus and Ava Helen Pauling, who was an undergraduate at Reed College in Portland in the mid-1950s. After graduation, she crossed the Atlantic to be near her brother, Peter, in Cambridge.

Pauling, Linus A native of Oregon (b. 1901), a professor in and chairman of the Chemistry Division at Caltech. As a senior at Oregon State Agricultural College, he met 18-year-old Ava Helen Miller, whom he married in 1923 after finishing his first year as a graduate student at Caltech.

Pauling, Peter Born in 1931, son of Linus and Ava Helen Pauling. After graduating from Caltech in physics and chemistry, he went in the fall of 1952 to the Cavendish Laboratory in Cambridge to work for a Ph.D. with John Kendrew as his supervisor.

Perutz, Max Austrian-British chemist (b. 1914). After graduating in Vienna, he went to Cambridge in 1936 to become a research student in X-ray crystallography of J. D. Bernal. In 1939, he came under the patronage of Lawrence Bragg, who was excited by Perutz's objective of using X-rays to solve the structure of hemoglobin. Through Bragg's backing, he became head, in 1947, of the Medical Research Council Unit for the Study of the Molecular Structure of Biological Systems.

Pontecorvo, Guido Italian-British geneticist born in Pisa (1907) whose brothers included the film director Gillo (*Battle of Algiers*), and the clever physicist Bruno (who left England in 1950 for the USSR when suspicion of his disloyalty arose). Originally an animal breeder in Tuscany, Guido moved to the Institute of Animal Genetics in Edinburgh, where, in 1938, he met the geneticist H. J. Muller and changed courses to pursue a Ph.D. under his supervision. He moved to Glasgow, becoming a Reader in Genetics in 1952 and later Professor.

Rich, Alexander Harvard-educated American physician-turned-biochemist (b. 1924), who moved to Caltech in 1949 as a postdoctoral fellow attached to Linus Pauling's laboratory. In 1952, he married Jane King, educated at Sarah Lawrence College, Bronxville, and the daughter of a long-established New York family.

Robertson, Mariette The daughter (b. 1932) of astrophysicist H. P. Robertson; she grew up in Pasadena and graduated from Wellesley College in 1953.

Rothschild, (Nathaniel Mayer) Victor A scion of the British branch of the famous banking family (b. 1910); educated at Trinity College, Cambridge. After heroic duties destroying aerial bombs during World War II, he returned to Cambridge as a member of its Zoology Department. In 1946, he married his second wife, Tess (Teresa) Mayor.

Sanger, Frederick Cambridge-educated English biochemist (b. 1918), whose subsequent work on proteins in Cambridge elucidated the amino sequence of insulin. For this work, he received the 1958 Nobel Prize for Chemistry.

Schramm, Gerhard German biochemist (b. 1910) who initiated work on tobacco mosaic virus at a Kaiser Wilhelm Institute in Berlin. During World War II he moved to Tübingen, where later he helped form the Max Planck Institute for Virus Research.

Schutt, Margot As a history student at Vassar College, she went to Edinburgh in 1952 for her junior year abroad. Returning to the U.S. on the S.S. *Georgic,* she and JDW became friends.

Simmons, Norman American biochemist (b. 1915) studying tobacco mosaic virus at the University of California in Los Angeles in the 1950s.

Stahl, Franklin Harvard-educated molecular biologist (b. 1929) who moved to Rochester for Ph.D. research under A. H. Doermann's supervision. In the summer of 1954, at Woods Hole, he met Matt Meselson, initiating a friendship that led to his moving to Caltech as a postdoctoral fellow in 1956.

Stanley, Wendell American chemist (b. 1904) who in 1935, at the then Rockefeller Institute Laboratory in Princeton, crystallized tobacco mosaic virus. For this accomplishment, he won the 1946 Nobel Prize for Chemistry (sharing it with John Northrop and James Sumner). In 1948, he became the first director of the new Virus Laboratory at the University of California, Berkeley.

Stent, Gunther Berlin-born (1902) physical chemist who trained at the University of Illinois and became a postdoctoral fellow at Caltech in the fall of 1948, joining Max Delbrück's group. After further years as a postdoc in Copenhagen and Paris, he joined Wendell Stanley's Virus Laboratory in Berkeley. In 1952, he married a young Icelander, Inga Loftdottir, then studying piano in Copenhagen.

Stewart, Janet Undergraduate (b. 1936) at Girton College, Cambridge in the mid-1950s and friend of Peter Pauling.

Stoker, Michael English virologist (b. 1916); studied medicine at Cambridge. After wartime service in India he returned to Cambridge and to research on animal viruses in the Pathology Department; a fellow of Clare College where he served as Medical Tutor.

Szent-Györgyi, Albert Hungarian-born (1893) biochemist who received the 1937 Nobel Prize for Physiology or Medicine for his isolation of vitamin C. In 1947, he moved to the U.S. where he established his Institute for Muscle Research at Woods Hole. While still in Hungary, he married his second wife, Marta.

Szent-Györgyi, Andrew Younger relation (b. 1926) of Albert Szent-Györgyi; researched with his wife, Eve, at Woods Hole in the mid-1950s.

Szilard, Leo Hungarian-born (1898) physicist who trained in Berlin (Ph.D. 1922), where he taught physics and was associated with Albert Einstein until Hitler came to power. He fled first to England and then to the U.S., where, with Enrico Fermi, he built the first nuclear reactor at the University of Chicago. After World War II, he held a professorship at Chicago's Institute of Radiobiology and Biophysics.

Teller, Edward Hungarian-born (1908), German-trained physicist (Ph.D. Leipzig 1930) who joined George Gamow teaching physics at George Washington University between 1935 and 1941. After World War II, he was professor at the University of Chicago before moving in 1953 to Berkeley.

Tissières, Alfred Medically qualified Swiss alpinist (b. 1917), who went to Cambridge after World War II to study biochemistry in David Kellin's laboratory. He was a prize fellow of King's College in 1951, and then went to Caltech for two years where he became a close friend of Max and Manny Delbrück. After returning to Cambridge, he resumed work at the Molteno Institute on oxidative phosphorylation with William Slater.

Todd, Alexander Scottish chemist (born in Glasgow, 1907); Professor of Organic Chemistry at the University of Cambridge from 1944, of towering presence (6 feet 4 inches tall). He and his research group established the covalent structure of the DNA and RNA backbones, receiving the Nobel Prize for Chemistry in 1957. In 1937, he married Alison Dale, a daughter of the noted physiologist Sir Henry Dale who earlier (1936) had won the Nobel Prize for Physiology or Medicine.

Wald, George Brooklyn-born (1906) and New York educated (Ph.D. Columbia University), he joined the Harvard faculty in 1935, becoming Professor of Biology in 1948. There he focused on the role of vitamin A in vision for which research he shared the 1967 Nobel Prize for Physiology or Medicine.

Wakefield, Lee Vassar College undergraduate who met JDW on the S.S. *Georgic* on its 1953 late-summer voyage from Southampton to New York.

Watson, Elizabeth ("Betty") Sister of Jim (b. 1930) who graduated from the University of Chicago in 1949. In 1951, she spent almost two years in Copenhagen and was in Cambridge during the months surrounding the discovery of the double helix.

Watson, James Born in Chicago (1928), educated at the University of Chicago and Indiana University (Ph.D. 1950). After spending a postdoctoral year in Copenhagen, he transferred in October 1951 to the Cavendish Laboratory at Cambridge University. There he met Francis Crick and together they found the double helix on February 28, 1953.

Weigle, Jean Jacques Swiss physicist-turned-phage-biologist (b. 1901) whose family wealth let him spend six months of each year at Caltech with Max Delbrück's group and six months at the University of Geneva, where earlier he had been chairman of its physics department. An experienced alpinist, he climbed with Alfred Tissières.

Wilkins, Maurice New Zealand–born (1916), he was educated at St. John's College, Cambridge, and studied for his Ph.D. (1940) at the University of Birmingham under J. T. Randall's supervision. After wartime research on uranium isotopes, he moved into biophysics first at St. Andrew's University and then at King's College London as a member of the newly formed Medical Research Council Biophysics Research Unit. Using DNA made in Switzerland, he and a student, Raymond Gosling, obtained an X-ray photo in 1950 of crystalline ("A" form) DNA.

Williams, Robley American electron microscopist (b. 1908) who joined Wendell Stanley's Virus Laboratory in Berkeley in 1950.

Wyman, Jeffries American biophysicist (b. 1901) of many Harvard connections, who became Scientific Attaché at the American Embassy in Paris in 1951.

Ycas, Martynas American biochemist (b. 1917). Worked for the Quartermaster Corps Laboratory, and then joined Syracuse

Editor's Note

The marginal notation indicates a moment in the narrative which corresponds to a letter among the ones collected in the appendix, "George Gamow Memorabilia."

Genes, Girls, and Gamow

Prologue

INITIALLY, WHEN I went back to Cambridge, no one would act as if I had been away. My breakfast over *The Times* at The Whim in Trinity Street would be the same—bacon, eggs, and toast. Only my hesitancy over the price would reveal that I then lived elsewhere. But when I returned in mid-September 1986, of my past only the buildings remained. The college porters no longer knew my face, and I had to identify myself to explain why I was in Clare College's garden along the Cam at a time when visitors who can read know they are trespassing.

I was back in Cambridge to look over a scientist whom I might want to recruit to my lab in the States. We were to meet with his research group for dinner at a French restaurant near Magdalene Bridge. I had time to spare and walked back over Clare Bridge through King's, listened to a trace of Evensong, and then was in Free School Lane and into the entrance of what was once the legendary Cavendish Laboratory. Much of modern physics had been discovered within its walls in the first third of the twentieth century, and it was there that I had come as a young American in the early 1950s.

Now the site had new inhabitants, the Department of Applied Biology—only a vague name in my memory and whose duties I had not a trace of recollection. The 6 p.m. bells at St Mary's Church had already rung, and I feared that the door would be bolted. But it was not, and I scampered up the stairs of the 1930s Austin Wing to the first-floor corridor where Francis Crick and I had shared an office. No plaque marked the birthplace of the double helix, and I opened the door expecting to find our former room deserted. Instead there, where Jerry Donohue and Peter Pauling once made four of us, was a solitary research student. He was using calipers to measure the dimensions of a sample of potatoes. Politeness cautioned me not to ask him why he had chosen this thesis topic. Instead I explained the abruptness of my interruption by telling him I had once also worked in this room and was curious how it was now used. By his response I realized he had no idea of who I was nor the intellectual frenzy that once dominated this utilitarian brick-walled space. The manners that Cambridge so long ago instilled in me did not let me reveal my identity, and quickly I was down the stairs and off to "The Golden Helix," the house where Francis and his wife, Odile, lived before their departure for California.

I walked to Market Street, into Sydney Street and then along Bridge Street on to Portugal Place. There, behind a narrow strip of pavement, were Nos. 19 and 20, the pair of Victorian row (terrace) houses in which the Cricks had lived for more than a quarter of a century. Originally they owned only No. 19, but after the fame that came with the double helix, they bought up the adjacent house to the south. That gave them the extra space for a yearly string of au-pair girls to look after their young daughters, Gabrielle and Jacqueline, whose youthful presence lightened the gravity of Francis's search for the genetic code. The houses were several hundred feet from the Moss Bros. shop, where, long ago, I had rented the formal clothes needed for college feasts.

As I looked down into the windows of the darkened basement dining room, I thought back on past pleasures from countless evenings of Odile's good food and Francis's spirited gossip about mutual Cambridge friends. Now, however, The Golden Helix and the tiny triangular garden in its front were ghostly quiet. Some years before Francis and Odile had begun to live more in southern California, at La Jolla with its Salk Institute, than in Cambridge, to which they returned only for the summer. First their cottage outside Cambridge was sold, and, then, just two weeks previously, The Golden Helix itself went to a Cambridge scientist returning from several years at Stanford University. With no one around, I continued to stare at the meter-long metal helix that Francis had attached at first-floor level. Seeing it, first-time visitors were reassured that they had found the right pair of houses.

I could only feel sad, not because the intellectual era that Francis had dominated was over but because Cambridge seemed not to care. I knew long, long ago that my time there was up, but for Francis to feel similarly was much harder to accept. Conceivably over 25 years at Cambridge was bound to yield boredom and the warmth and blue skies of California must have added to other reasons for moving. But would this path have been so inevitable if the university had not been so immersed in its history that its institutions were always more important to it than its inhabitants. Only the towering Glaswegian chemist Lord Todd could assert his godliness and get away with it.

Francis was too much like a fast bowler from the West Indies—batting members of the team did not stand a chance. With him in the room, you could never have a moment of relief from a succession of quickly thrown ideas. Even as he reached 60, he just would not show his age. It was not on the cards for him to retire physically—much less intellectually—at 65, and making him the master of his college would have been a grave misuse of his talents. Something had to move and it was not Cambridge.

1

Cambridge (England): April 1953

ALTHOUGH MY HAIR was properly long and my accent toned to suggest almost an English origin, Odile Crick told me I had still far to go before I would look right walking along Cambridge's King's Parade, much less looking purposefully indolent in one of its college gardens. My appearance would not have mattered if I were the same as a month ago—an unkempt slender figure who said what I thought as opposed to what good manners required. But now that Francis Crick and I had given the world the double helix, Cambridge in its own quiet way was bound to ask what we looked like. The time had come to acquire at least one set of clothes that would go well with Francis's Edwardian elegance. I was not trusted to act alone and Odile accompanied me to the men's clothing shop across from the chapel of John's (the College). My ill-fitting American tweed jacket was thrown out and replaced by a blue blazer and associated gray trousers. They would much better express my new status as the co-winner of a very great scientific jackpot.

The DNA molecule we had found two months before—in March 1953—was far more beautiful than we ever anticipated. With the two polynucleotide chains held together by adenine-thymine and guanine-cytosine base pairs, DNA had the complementary structure needed for the gene to be exactly copied during chromosome replication. When 1953 started, finding out what genes look like and how they replicate were two of the three big unsolved problems in genetics. Seemingly coming from nowhere, Francis and I had now grasped both. At times I

7

virtually had to pinch myself to prove that I was not in the middle of a wonderful dream. But I was not, and so the possibility existed of a grand slam in which Francis and I also worked out how genes provide the information to make proteins.

By the flip of a coin, our names in the original manuscript had the order Watson-Crick instead of Crick-Watson. So several Cambridge wags now could refer to our DNA model as the WC structure. They suspected that our golden helix would be found tainted and destined for dumping down the water-closet drain.

I had become monomaniacal about DNA only in 1951 when I had just turned 23 and as a postdoctoral fellow was temporarily in Naples attending a small May meeting on biologically important macromolecules. There I learned from a mid-thirtyish English physicist called Maurice Wilkins that DNA, if properly prepared, diffracts X-rays as if it were a highly organized crystal. The odds were thus good that DNA molecules (genes) themselves have highly regular structures that conceivably could be worked out over the next several years. Briefly I considered asking Wilkins if he would let me join his London lab at King's College on the Strand, but my attempts to talk with him after his lecture elicited no enthusiastic response and I dropped the idea.

Instead, through the intervention of Salvador Luria, my Ph.D. supervisor at Indiana University, I was taken on five months later at the Cavendish Laboratory in Cambridge to work with an English chemist, John Kendrew. He was helping the Austrian-born chemist Max Perutz lead a small research group supported by the Medical Research Council (MRC) called the "Unit for the Study of the Molecular Structure of Biological Systems." Started in 1947, its scientists used X-ray methods to work on the three-dimensional structures of the oxygen-carrying proteins hemoglobin and myoglobin. In going to join the group, I hoped to expand the attention of the unit to DNA, so that they would let me work on it, instead of a protein, once I had learned X-ray diffraction techniques.

My crystallographic career, however, would have likely soon aborted if Francis Crick had not been in the lab. From the moment I arrived, he treated me as if I was a much younger brother in need of help. Then 35 years old, Francis was effectively unknown outside Cambridge, having

Cavendish Laboratory group photo, spring 1952. In the center of the first row is Sir Lawrence Bragg; JDW is second row, sixth from left, next to Hugh Huxley and Francis Crick.

joined the unit only two years before. Already Francis's penchant for theory had made him a powerful addition to the team's protein-solving efforts. His first major success came soon after I arrived, when that October he helped work out the theory for diffraction from helical objects. Even so, Francis could not anticipate a long-term future within the unit, because the week before he had badly upset the head of the Cavendish Laboratory, Sir Lawrence Bragg, by arguing that he, not Bragg, first saw a potential new way of analyzing protein X-ray diffraction patterns. To say the least, Bragg did not like the implication that he had pinched a younger colleague's idea. In fact, on that ill-fated Saturday morning, Francis realized that neither his nor Bragg's precise approaches were that good and that only isomorphous replacement methods held out real hope.

That fall of 1951 we had no reason to hope that we would be more than minor players in DNA research. The experimentalists at King's College London—Maurice Wilkins and Rosalind Franklin—were set to provide the definitive evidence for choosing one DNA model over another. But over the next year, their personalities clashed badly, and Maurice found himself driven away from X-ray analysis of DNA. Soon Rosalind had all the cards needed to solve the structure, provided she co-opted the model-building approach that Francis and I so passionately argued for.

Here her greatest mistake was being put off by Francis's strong personality that she thought masked a bumptious overextended intellect.

Even less predictable was the inexplicable chemical botch that Linus Pauling, then universally perceived as the world's best chemist, made with his ill-conceived triple-stranded DNA helix. Late in 1952, we had become apprehensive when Linus's son, Peter, who had newly arrived in the unit to be a research student with John Kendrew, told us that "Pop" was working on DNA. Only 18 months before Linus had humiliated the Cambridge group with his α-helical fold for proteins. We breathed much, much easier in February 1953 when we read a manuscript from the California Institute of Technology (Caltech) and saw that Pauling's DNA model was way off the mark.

Quickly I raced into London to alert the King's group that Pauling's new helix was a botch and we should expect him quickly to devise a better model. Rosalind, however, thought I was being unnecessarily hysterical, telling me in no uncertain terms that DNA was not helical. Afterwards, in the safety of his office, Maurice—bristling with anger at having been shackled now for almost two years by Rosalind's intransigence—let loose the, until then, closely guarded King's secret that DNA existed in a paracrystalline (B) form as well as a crystalline (A) form. In his mind the cross-shaped B-diffraction pattern, shown on the X-ray he then impulsively took out of a drawer for me to see, had to arise from helical symmetry.

Almost perversely, it was Linus Pauling's entry into the DNA game that made it possible for Francis and me to find the double helix. In November 1951, before it was clear that Pauling was out to get the DNA prize, Francis and I had been told by Sir Lawrence Bragg that DNA was off limits to the Cambridge unit because it belonged to the workers at King's. Even 14 months later, bad memories still existed of our awkward first attempts to build DNA models. But we then quickly gave up trying to guess the DNA structure and even passed details of the molds needed to build models to Maurice Wilkins. By now appraised of the B-form's existence, Bragg wanted Francis and me to have another go at building models. He hoped that our efforts—possibly coordinated with those in London—would generate the right answer before Pauling recovered his senses.

No one then could have anticipated that in less than a month Francis and I alone would have found the answer and one so perfect that the experimental evidence in its favor from King's almost seemed an unnecessary accompaniment to a graceful composition put together in heaven. Our writing of the tiny manuscript for *Nature* that would announce the double helix seemed even then an historic occasion. My sister Elizabeth, who had followed me to Europe two years before, did the typing, with Odile Crick using her artistic talents to draw the intertwined, base-paired, polynucleotide chains. Together with two experimental manuscripts from the warring King's groups of Wilkins and Franklin, it was dispatched to *Nature*'s editor by Bragg on April 2 and published only slightly more than three weeks later on April 25.

Our most unanticipated success was a big relief to Betty, my sister, and Odile. My proclivity for super dreams had clearly long worried Betty, who feared that I would never adapt successfully to the world of ordinary people. Odile, on the other hand, no longer had to worry about having to leave Cambridge. Bragg could not force Francis to leave the lab after he had helped give England the double helix. And, even though it was then ordained that the Cricks go for a year to Brooklyn, Odile would not have to consider the awful fate of staying on.

Dinners with the Cricks at their house in Portugal Place became even more spirited occasions after our success, with Odile often bantering me about my better prospects of getting the perfect girlfriend. Before the double helix, it was easy to meet the foreign girls who were in Cambridge to learn English. But I sensed it would be much better to try and get to know the undergraduate English girls at Girton or Newnham—at least I might understand what they said. But no one I knew then had any real contacts at these women's colleges. The correct tack for me might have been to seek out an attractive tennis player. But Francis, though his father played at Wimbledon, had long ago given up outdoor sports and neither he nor Odile knew any girl, blond or otherwise, who hit the ball hard. Happily by the time of our discovery, and on my own initiative, I thought I might have located the girlfriend appropriate for my new fame.

The previous August, in the Italian Alps, I had met a good-looking English girl called Sheila Griffiths, who was living with a mountaineering family. As luck would have it, we started talking only two days before I was due to depart, one of which she spent ascending Monte Disgrázia that loomed above the tiny village of Chiarregio. Born in Wales, Sheila was in Italy to improve her Italian in return for looking after two children and hoped to go to Rome when the summer ended. She came from a mining heritage and her father, Jim Griffiths, was a Labour Member of Parliament. She had several more weeks in the mountains and worried about keeping busy if bad weather settled in. So I lent her my copy of Aldous Huxley's *Point Counter Point* and, when briefly in Milan, bought copies of *The Economist* and *New Statesman* to post on to her.

During the fall I kept hoping to hear from her, having given her my Clare College address when we parted because then she did not know where she would be living in Rome. Just before we found the double helix, however, she sent me a letter from the Dolomites where she was learning to ski with her two charges. At Easter she was coming permanently back to England and enclosed the telephone number of her family's home in Putney. Before we parted in Italy, I had told her that DNA had to be at the heart of life. Now, in April 1953, this was no longer a conjecture: the double helix would soon be a, if not *the,* fact of life.

Cambridge (England): April–May 1953

IN LONDON, SHEILA Griffiths and I first met in Mayfair at Brown's Hotel near the Society for Visiting Scientists in Old Burlington Street, where, for 17 shillings and sixpence, I got a bed to sleep on and corn flakes and toast for breakfast. Immediately I told her of our manuscript that would be appearing in *Nature* the next week and very likely create a big splash. Later, as we dined at the Dover Street Buttery, we had much conversational fun, and the evening went by far too fast. But I already knew of a date two weeks off when Alicia Markova was to dance *Giselle* at Covent Garden, and I had no difficulty persuading Sheila to join me for the occasion.

Several days earlier I had put my sister on the boat train to Southampton for her return to the States and to our parents, now living amongst the Indiana sand dunes. Betty had been in Europe for almost two years, from just before my first meeting with Maurice Wilkins. Initially we were easily spotted as Americans, especially me with my closely cropped hair and lumberjack shirts. But Betty had acquired a continental flavor from her Jacques Faith suits and when I spoke I was no longer recognized as an American. To my surprise, I often passed as Irish, possibly reflecting the language of my Gleason grandmother, who lived with my family when I was growing up. This expatriate phase of our lives, however, was soon to end; she would have to stop being called Elizabeth and be Betty again—a necessary transition from the English to the American way.

13

Nevertheless Betty looked forward to going home more than I did. In the late summer she would be setting off again, this time to Japan, to marry an American whom she had known at the University of Chicago. Likewise I was to return to the States at summer's end to take up a post-doctoral position at the California Institute of Technology in Pasadena. Although I loved my Cambridge life, I saw no way of delaying my departure. For almost a year before the double helix was found, my longtime patron, Max Delbrück, had been counting on me to come to Pasadena to help with the students who were working with viruses that infect bacteria—bacteriophages, or phages, for short.

Before April had ended, Crick and I had dispatched a second paper to *Nature* to elaborate on the phrase, "It has not escaped our notice that the specific pairing we have postulated immediately suggests a possible copying mechanism for the genetic material." Francis initially had wanted to be much more specific in our April 25 paper, but I argued that we should understate our model's implication because our paper was to be followed by ones from Rosalind Franklin's and Maurice Wilkins's groups, the two having long gone their separate ways. But once our manuscript had gone to *Nature,* I, too, worried that if we didn't state our ideas more clearly somebody else would try to poach them and get some of the credit. Francis wrote most of this second manuscript, which we called "Genetical implications of the structure of deoxyribonucleic acid." We had less than a week to complete it and as soon we had finished the drawings it went off with Sir Leonard Bragg's imprint to appear in the May 30 issue of *Nature.*

All through those spring days, visitors constantly popped in to see the model. Dorothy Hodgkin, England's best crystallographer, came over from Oxford with her postdoc student Jack Dunitz, as did later the young theoretical chemist Leslie Orgel, then at Magdalen College. Leslie brought along the short and almost-pudgy Sydney Brenner, who had finished his medical degree in South Africa two years earlier. Sydney, then 26, was attached to the lab of the well-known Oxford physical chemist Cyril Hinshelwood, long notorious in genetics circles for his Lamarckian interpretation of bacterial heredity. But in Johannesburg, Sydney had no way of knowing this about Hinshelwood. Later, Hinshelwood was to get the Nobel Prize for Chemistry as well as become President of the Royal Society.

Upon arriving in Oxford, Sydney quickly came to his senses and chose to do his Ph.D. thesis research on phages, about which his professor knew close to nothing. By the time we met, Sydney knew that his own phage work would not make a noticeable splash but common sense required him to see his experiments through to their unimportant end. Then he could move on to more promising research objectives. Knowing of Francis's sales pitch about DNA even before he spoke, Sydney quickly retreated out of earshot from the powerful Crick voice, and we began a four-hour-long, non-stop conversation on the possible involvement of ribonucleic acid (RNA) in protein synthesis. Later, when we came back to Francis's presence, I felt more than itchy. Like everyone else now within range of Francis's booming enthusiasm, I was becoming allergic to DNA. So I could not help telling others later that I would barely survive another tour of the base pairs. Learning of my possible defection, Francis took me aside and told me I didn't realize our work's deep significance.

So warned, I even more anticipated going down to London to see *Giselle* at Covent Garden with Sheila Griffiths. My American stipend of almost a thousand pounds let me get very good stall seats for close viewing of Markova, who I had not seen dance before. During the performance we went through a tiny box of Sheila's chocolates and afterwards sought out more coffee and conversation before catching our respective trains back to Putney and Cambridge. Then I learned that she had asked a friend to get for her the April *Nature* issue that she believed contained our article. But she wondered why it made no mention of DNA. Embarrassingly, I realized that she had been given the April 18 issue, not the April 25 issue that included our DNA work.

The previous September, in the midst of a wine-filled lunch above Lake Locarno, the Paris-based geneticist Boris Ephrussi and I, with his Swiss postdoc Urs Leopold, composed a silly note about terminology in bacterial genetics. We wrote it as a spoof of the turgid writings of Joshua Lederberg, justly much famed for his 1946 discovery at Yale University that bacteria genetically recombine. Later we got our Swiss physicist-turned-biologist friend, Jean Weigle, to add his Geneva address, and dispatched the effort to *Nature* to see if we could trick its editor into publishing it. Initially I was delighted when a postcard from the editor told us that our inane ramblings would be printed. Later, however, I became apprehensive that they would simultaneously appear with the real thing. So I was more than relieved when it came out the week before. Sheila's tone now expressed doubts whether our April 25 bombshell even existed. Still, she clearly wanted to see me after the trip I was soon to make to Scotland.

In Edinburgh, I stayed with C. H. Waddington, the postwar Professor of Animal Genetics, who lived in a Roman-style villa that had been built in the early nineteenth century as the city expanded to the south. Talking to Waddington was a bad letdown, since I had long respected him because of his excellent 1939 book on modern genetics that I used as a graduate student at Indiana University. To my surprise, he now seemed indifferent to the double helix, appearing refractory to the idea that DNA's complementary structure was at the heart of the copying mechanism for genes. Why he was so dense eluded me then. Only later did I realize that Waddington wanted something more important than simple molecules to control the key attributes of living organisms. In contrast, in Glasgow, its Genetics professor Guido Pontecorvo instantly followed my argument. The next day we took a walk towards Loch Lomond so that we could be served alcohol, because, on Sundays, publicans could legally serve drinks only to persons living more than five miles distant.

After my return to Cambridge, I wore my new blue blazer when Francis and I were photographed for *Varsity*, the undergraduate paper that came out twice a week during term. A news story was being written about our Cavendish breakthrough and their main photographer, A. C. Barrington-Brown, spent a morning with us. The occasion was inherently jolly, with Francis making sure that our picture-taker not only got our names right but also learned why our two-stranded model would revolutionize biology. Lacking Francis's English polish, my attempt to look serious next to the base pairs led to several photos marked by silly grimaces on my face. I looked more acceptable when photographed next

(Reprinted from Nature, Vol. 171, p. 701, April 18, 1953)

Terminology in Bacterial Genetics

THE increasing complexity of bacterial genetics is illustrated by several recent letters in *Nature*[1]. What seems to us a rather chaotic growth in technical vocabulary has followed these experimental developments. This may result not infrequently in prolix and cavil publications, and important investigations may thus become unintelligible to the non-specialist. For example, the terms bacterial 'transformation', 'induction' and 'transduction' have all been used for describing aspects of a single phenomenon, namely, 'sexual recombination' in bacteria[2]. (Even the word 'infection' has found its way into reviews on this subject.) As a solution to this confusing situation, we would like to suggest the use of the term 'inter-bacterial information' to replace those above. It does not imply necessarily the transfer of material substances, and recognizes the possible future importance of cybernetics at the bacterial level.

BORIS EPHRUSSI

Laboratoire de Génétique,
Université de Paris.

URS LEOPOLD

Zurich.

J. D. WATSON

Clare College,
Cambridge.

J. J. WEIGLE

Institut de Physique,
Université de Genève.

[1] Lederberg, J., and Tatum, E. L., *Nature*, **158**, 558 (1946). Cavalli, L. L., and Heslot, H., *Nature*, **164**, 1058 (1949). Hayes, W., *Nature*, **169**, 118 (1952).
[2] Lindegren, C. C., *Zlb. Bakt.*, Abt. II, **92**, 40 (1935).

JDW and Guido Pontecorvo in the Alps above Saas Fee (August 1953)

to Francis at his desk and sent a print to my parents confirming the news
that indeed we were onto something more than important.

During those mid-May days of 1953, I repeatedly failed to get through
to Sheila Griffiths—her Putney telephone number seemed perpetually
engaged. Thinking I might have better luck in London, I tried again
unsuccessfully to reach her after I went to see Rosalind Franklin, who
by then had moved to J. D. Bernal's lab at Birkbeck College in Torring-
ton Square. There Rosalind had been calculating the equatorial reflections expected from our model and was finding them not in agreement
with her measurements. Conceivably the radius at which we placed the
phosphate atoms in our double-helix model was not quite right.
Nonetheless, I felt a little nervous on the train back from the Liverpool
Street Station—a Watson-Crick folly was sure to be long remembered.
Yet could something so perfectly pretty really be wrong? Happily no one
of even half authority thought so, and later that week I happily watched
John Kendrew's wife, Elizabeth, making dramatic large sketches of the
model.

My quietly elated spirits were further boosted when I was invited at
the last minute to the forthcoming June symposium on viruses at Cold

Spring Harbor Laboratory, Long Island, New York. To my delight, Max Delbrück had written to the lab's director, Milislav Demerec, that I should there present the double helix. So Francis and I spent the last two weeks in May putting together our symposium paper, frequently arguing out the precise language. For the first time we discussed how spontaneous mutations might occur and less successfully I took on the question of how chromosomes paired. In the end, my ideas about crossing-over didn't gel, and we knocked out this topic from the final manuscript. During our writing, I finally heard from Sheila Griffiths who enclosed the copy of Huxley's *Point Counter Point* that I had loaned to her the previous summer. By then she knew that my talk about the double helix was no bluff for she mentioned reading Ritchie Calder's piece on it in the *News Chronicle*. But there was no time to see her before I flew to the States.

By the time the manuscript was finished, Tony Broad, who had come to Cambridge to build our rotating anode X-ray tube, had made the first demonstration model of the double helix. Even in its plastic case it was beautiful! So on June 1, I put it under my arm, went down to London, took the airport bus to Heathrow, and got on a BOAC Constellation. It flew to New York with almost no passengers, for the next day was the Coronation of Elizabeth II. We stopped at both Prestwick and Gander and the overnight flight lasted some 18 hours. In the morning, as we were nearing Long Island, the pilot used the plane's loudspeakers to say that Edmund Hillary and Sherpa Tenzing Norgay had reached the top of Mt. Everest on May 29.

Cold Spring Harbor: June 1953

AS THE PLANE approached New York along Long Island's Atlantic side, I eagerly looked to the north hoping to see the extensive grounds of the Cold Spring Harbor Laboratory. This tiny institution, located at the end of a long and beautiful harbor off Long Island Sound, still had a pastoral aura about it—perfect for chasing ideas at the measured pace that allowed time for afternoon tennis and much uninhibited dinner talk. That it was so intellectually high-powered was not apparent from its physical appearance. At first glance it resembled a deteriorating New England village, strung out along a country lane with many of the buildings dating from the days of a tiny whaling industry that barely thrived before the Civil War.

My first period at the lab was in 1948 when I accompanied my thesis adviser, Salvador Luria, from Indiana to join the still-small phage group there. The group's members, for the most part physicists or chemists, then were revolutionizing genetics through their studies on bacteria and their viruses, the phages. Luria, a Jewish refugee from Fascist Italy, was a familiar face at Cold Spring Harbor, having first gone there in June 1941 for an important meeting on genes and chromosomes. When it finished, he stayed on to do phage experiments with his newly acquired theoretical physicist friend, the Protestant German Max Delbrück, who then saw no future in Hitler's Germany.

Soon after my arrival, Max, his wife Manny, and their two young children had already wearily settled into their apartment after a night flight

Cold Spring Harbor Laboratory, Long Island, from the inner harbor

from Los Angeles. They had married in 1941. Max, born in 1906, was 12 years Manny's senior but now they did not look different in age, as Max, who was six feet plus, had retained the slenderness of his youth. Increasingly, during my first Cold Spring Harbor summer, I had wanted to be as much as possible like Max, even to having a wife like Manny, whose Scottish good looks went well with a free-spirited mind that sought out friends with novel backgrounds and viewpoints. Most appealing were her love of the outdoors, the determined way she played a good game of tennis, and her dislike of academic couples who derived no fun from good-natured mischief.

By that evening, virtually all the world's key scientists who worked with phages had arrived and after dinner we began talking either in small groups standing in front of Blackford Hall where we ate or sitting around the big wooden tables on the east-facing porch that faced the harbor. The scope of this Cold Spring Harbor symposium was wider

Manny Delbrück and JDW at Cold Spring Harbor, June 1953

than phages, however, and many leading researchers on plant and animal viruses were attending as well. That this 1953 symposium could be
so big, with some 270 attendees, reflected Max Delbrück's influence on
the National Foundation for Infantile Paralysis, an organization that
then saw the need to support research on bacterial viruses as well as
those that infect animals. Moreover, it provided the fellowship monies
that supported me at the time the double helix was found.

Such a large symposium could be held at Cold Spring Harbor only
because a new, Scandinavian-style auditorium had been completed a
few weeks before. Notwithstanding its hard green seats, this auditorium
was a great triumph for the Yugoslav-born Milislav Demerec, who had
been the Lab's director since 1941. Originally a classical plant geneticist,
for many years Demerec had been studying the famous fruit fly
Drosophila. Now he was moving much of the long-term research at Cold
Spring Harbor into the new era of microbial genetics, both through his
own work on bacterial heredity as well as by bringing to the Lab from
Washington University in St. Louis in 1950 the great talents of the
unusually taciturn chemist Alfred Hershey. Only 18 months later, Hershey with his assistant, Martha Chase, did their famous experiment

showing that DNA carries the genetic specificity of phages—a discovery that spurred me even more into finding out what DNA looked like in three dimensions.

Clearly wanting to talk with Al Hershey, and vice versa, was the legendary Leo Szilard, who after the war had decided to switch from physics into biology. In the 1920s, Hungarian-born Leo had first done physics in Berlin and later came as a refugee to the States, where, in 1942, he built with Enrico Fermi the first nuclear reactor at the University of Chicago. Afterwards, Szilard's desire to patent their accomplishments so greatly irritated the military authorities in control of the Manhattan Project, that General Groves attempted to jail him until the atomic bomb had been used successfully. Although Groves was overruled by Secretary of War Henry Stimson, Groves nevertheless made certain that Szilard never later joined the bomb-design group at Los Alamos.

Politically unsuccessful in his efforts to prevent the U.S. from using atomic bombs in Japan, Leo needed another big objective. So knowing of Max Delbrück's move into biology, he came to Cold Spring Harbor in the summer of 1947 to take the two-week-long practical course that Max had started two summers previously to attract more scientists into phage work. Invariably dressed in white seersucker suits that had no chance of concealing his portly frame, Leo did much of his thinking during long bathtub hours. Equally long hours he spent eating, whenever possible joined by associates whose brains he could pick.

The first rows of symposium seats traditionally were occupied by those with the courage to interrupt speakers whose thoughts had gotten out of hand. Max and Leo were so positioned when Demerec opened the meeting by noting the large number of attendees and thanking the sponsors for providing the necessary travel funds. As soon as Demerec had finished, Szilard jumped up to thank Demerec for going to the great trouble of locating the wartime-like powdered eggs that were being served at breakfast (not since the war had such a delicacy been available to accompany the morning papers). In a more serious vein, Delbrück then went on with an overview of the intellectual food that was to sustain us for the next seven days, noting that the program was arranged around the life cycle of a virus, starting from its free state, through its

At the Cold Spring Harbor Symposium, June 1953: (from left to right) Max Delbrück, Aaron Novick, Leo Szilard, and JDW

infection, and subsequent multiplication and release from its respective host cell.

In the past, Max had ridiculed the idea that viruses could exist within certain cells in a latent proviral form distinct from the vegetative phase. But at last summer's phage meeting at the Abbaye Royamont north of Paris, Max capitulated to André Lwoff, a French microbiologist, after learning of his definitive experiment done at the Institut Pasteur in Paris. Increasingly by then, I and other younger members of the phage group were realizing that Max's scientific hunches were frequently dreadful. But once he saw the truth he quickly and gracefully reversed course, as shown by his introductory symposium homage to the intellectually penetrating Lwoff, who that day also sat in one of the front rows. Later Lwoff, who was of Russian-Polish extraction, saw no reason himself to present a paper because that was to be done in part by his younger colleague, François Jacob. The new intellectual heavyweight on the scene, François would sit quietly for hours, then suddenly come to life with questions on what the experiments actually told us.

I gave my report on the double helix halfway through the meeting. There was almost no reason to talk because most participants had already seen my foot-long demonstration model with the A–T and G–C base pairs colored red and green, respectively. Moreover, before the meeting, Max had distributed copies of the first trio of *Nature* papers to all participants. So my talk could be very brief. I gave it with my loose shirt overlaying short trousers and tennis shoes unencumbered by laces. Not a trace of Odile Crick's attempt to make me look English remained and, temporarily, it was as if I had never been to England. In emphasizing Francis's key role, only the plant virologist Roy Markham, the sole other attendee from Cambridge, knew that I was not exaggerating the Crick intellect nor personality.

With the double helix so perfectly the answer to the dreams of geneticists, little questioning occurred immediately following my talk. But afterwards Barry Commoner, the Washington University biologist, came up to disagree, not wanting to believe that his DNA dye-binding studies of three years before were hopelessly wide of the mark. Leo Szilard's concerns were of a different sort—shouldn't I patent the double helix? But he knew that his proposal would not fly and told me I could only be famous, not rich.

I found the meeting ending much too soon, with phage friends from the past soon departing home to their wives, but I knew the Delbrücks were to remain at Cold Spring Harbor for the summer. Even before my arrival I planned to stay on at the Lab for another week before returning to England. Later I decided to extend my visit for a further week when I found out that the ornithologist Ernst Mayr's 17-year-old daughter Christa would be spending some time in Cold Spring Harbor once she graduated from high school. The Mayrs had spent the previous 10 summers at Cold Spring Harbor, coming out from their base in New York, where Ernst worked at the American Museum of Natural History. Born and educated in Germany, Ernst had arrived in America in 1932 after spending several years collecting birds in the Solomon Islands on a bird expedition sponsored by the British philanthropist Walter Rothschild. Ernst had just been appointed Professor of Zoology at Harvard University and soon he and his family would be relocating to the American

Ernst Mayr on Blackford Porch, Cold Spring Harbor Lab, 1950

Cambridge. Before I went to Europe three years earlier, my eyes used to turn first to Christa's year-younger blond sister, Susie, adorably pretty since birth. Now, however, it was the more intellectual, 17-year-old, brown-haired Christa that I found myself anticipating.

By the time Christa had arrived, the summer's phage course had started, and one of the students was Bathsheba de Rothschild from Paris. She was excited about the ideas that Leo Szilard could bring to biology and was working in a lab at Columbia. Bathsheba was keen on modern dance and a major patroness of Martha Graham, whom she brought out for Sunday lunch on the Blackford porch. In the evenings, I accompa-

Christa Mayr in Cambridge, Massachusetts, 1954

nied the Mayrs as they sought their good-natured after-supper excitement. One evening we went into nearby Huntington for a Hollywood movie and, afterwards, joined a communal square dance in a field off Highway 25A. For the first time I was swinging Christa as a full-bodied woman, no longer the gangly child of earlier memories. Not at all shy and calmly confident of her beliefs, she was bubbly excited about being accepted by the academically prestigious Swarthmore College.

The night before my flight back to England, we were again part of a square on the big lawn underneath the Charles Adams–like Carnegie dorm that long had been home for unmarried women working as assistants to the more senior scientists. I no longer felt awkward at our changed relationship. After the dancing ended, Christa and I walked down Bungtown Road to the sandspit where we sat next to each other on the warm beach sand, recalling shared memories from past summers. Increasingly I wanted to touch her but, fearing a negative reaction, avoided even holding her hand as we walked back to the Lab's center. By then, the sun was just rising and Blackford Hall was starkly beautiful in its 1906 classical Italianate form. With butterflies rumbling through my stomach and not wanting to awaken her parents, we silently bid each other adieu on the Mayrs' doorstep. Christa then slipped through the door and I went up to my monastically bare Blackford room.

Neither of us was more than half-awake when three hours later I came to say good-bye to her parents and Susie, and to Max and Manny Delbrück who lived in the next apartment. Soon a train journey would start my journey back to England. Until three weeks before, its fabled Cambridge represented all I wanted from life. Now in love, I knew otherwise.

4

Cambridge (England): July–August 1953

A CROWDED STRATOCRUISER took me back to England and Clare
College, where I was to remain only two months before moving to the
California Institute of Technology, or Caltech for short. After takeoff I
dozed off, coming back to life when we made the fueling stop in Gander.
The plane carried Krishna Menon, Nehru's well-known emissary, who
was shuttling back to Delhi from a stint at the United Nations. At Gander he made himself very visible as he walked up and down the long
waiting hall looking self-important. Later, when the plane landed in
London, Sarah Churchill got off first into a waiting Rolls. Winston, her
father, had had a stroke two nights before.

During my absence, the first newspaper article that reflected an
interview with Francis Crick came out. It was in the *Sunday Telegraph*
and reached a large audience. As soon as I saw it, I told Francis we
should not help generate more publicity. But he strongly disagreed, seeing no reason why the nature of our triumph should not be distributed
widely. To me, discretion demanded that we stay quiet and let history
judge us, not our peers—at least not minor Cambridge biochemists.
Later, when Francis was asked to give a talk on the BBC, I consented
only if it was broadcast exclusively on its overseas services. I was afraid
that we might be thought grabby and did not want to stir up more discussion as to whether we had improperly used the King's College data.

Christa Mayr was much on my mind, but initially I wanted to see
Sheila Griffiths again and wrote her a brief note saying that I was back

29

from the States for a brief visit to complete work on a manuscript about the double helix. With it I enclosed a copy of *Microcosmographica Academica,* a tiny analysis of how to succeed in Cambridge University politics, written in 1908 by Francis Cornford, then a Professor of Classics. Much of what he wrote still held true, and I thought Sheila would enjoy it, particularly because she would also be hearing about Cambridge from Roy Pryce, a young historian friend whom she met in Rome. She replied quickly, indicating her concern about why I had been so long in contacting her. She had spotted a second *News Chronicle* story about the double helix and wondered whether it was also big news in the States. The way still seemed open to let us know each other better. But now, caught up with love for the newly grown-up Christa, I prevaricated and the summer was to pass without our meeting.

Soon I was concentrating on a detailed manuscript for the *Proceedings of the Royal Society* on how we arrived at the double helix. This time I did virtually all the writing because Francis was finishing his thesis, soon to be forgotten, on the shape of hemoglobin. As I got near the end, I began to relax. It was the first time I had put together language of the type that Lawrence Bragg, Max Delbrück, and Linus Pauling had mastered so well. After it was ready for Bragg to communicate to the Royal Society, I spent a weekend with Sydney and May Brenner in Oxford. There, indifferent to their young son Jonathan's wishes, I ate all their chocolates, while urging Sydney to go to Cold Spring Harbor when he was finished at Oxford. May, however, was heavily committed to left-wing causes and did not warm to the idea, reacting strongly against the McCarthyism then rampant on the American scene.

By then I had received a short chatty letter from Christa in response to one I'd sent her soon after my return. The hope that she might reply had kept me checking daily for a letter in my pigeonhole at the Porter's Lodge. My first letter from the States, however, came from my sister, Betty, still with our parents preparing for her lengthy flight to Tokyo and marriage to Robert Myers. Before she left Cambridge, Peter Pauling urged her to spend a night in Honolulu where his brother, Linus, Jr., was living, married to the heiress Anita McCormick Rockefeller Oser. To Betty's delight, this stopover was arranged, and she could see whether Linus, Jr. was also a charming Pauling. A separate letter from my mother told me that Betty's wedding unfortunately had to occur in Tokyo rather than near Chicago, so our family friends would miss seeing who Betty was marrying. Mother, moreover, feared that Bob Myers might be a Republican and wondered why her daughter was going through a governmental security check. Betty was sure of her choice, though, and I no longer worried that she might marry someone of charm but no consequence.

For several weeks after my return, I kept open the possibility of submitting the double-helix research for a Cambridge Ph.D. Having two such degrees clearly made no sense except for future conversational gambits. But without the Ph.D. pretense, I would not now be ensconced on the R staircase of Clare College with windows overlooking the giant Atlas cedar inside Memorial Court. Although much of the thesis could consist of published material, I would have to write a proper introduction that with perseverance should only take up two or three weeks. By midsummer, however, I saw myself running out of time and semi-reluctantly abandoned thought of the splendid red academic robe worn by holders of Cambridge Ph.D.'s. I equally saw that it made no sense to pursue a research fellowship at Clare. Michael Stoker, its medical tutor, thought that, given the double helix, I would most certainly be awarded one if I applied. But despite its exquisite architecture and perfect garden, I sensed that Clare was unlikely to provide me with a viable social life.

At our lunches in the Eagle, the nearby pub on Bene't Street where we invariably had lunch, Francis and I increasingly batted about how the genetic information encoded by the sequence of base pairs in DNA might be used to determine the order of the different amino acids in proteins. Here the DNA base pairs could not directly provide template surfaces attracting specific amino acids because clean experiments had shown that amino acids assemble into proteins in the cytoplasm, totally segregated by nuclear membranes from the chromosomally located DNA. For over a year I had been telling Francis that the genetic information of DNA chains must first be copied into that of complementary RNA molecules. These then must move to the cell's cytoplasm to function as the templates ordering the amino acids in proteins.

But until we found the base pairs, I had no idea at the molecular level about how the genetic information of DNA could be passed into

RNA molecules. Now the answer was obvious. The same base-pairing scheme involved in copying DNA could be used to make single-stranded RNA molecules upon complementary single-stranded DNA molecules. Of course, there was no firm evidence to convince skeptics that this DNA→RNA→ protein scheme actually existed. But if RNA wasn't the template for protein synthesis, why should there be so much in cells, particularly those involved in extensive protein synthesis, such as liver cells? Most certainly RNA was not there, say, to control the viscosity of cells, which Francis measured for his first two years in Cambridge at the Strangeways Laboratory as his introduction to biology. Cracking the RNA structure was my next main task. With luck, I could quickly work it out once I got to Caltech.

Even, however, without knowing the RNA structure, we knew that each amino acid must be selected by groups of base pairs. There were far too many amino acids for a one-to-one correspondence. At first glance, the number of known amino acids was more than 25 whereas there are only four bases in the DNA alphabet. But most proteins have only a smaller subset of amino acids. Oddball amino acids, like the hydroxy-proline of collagen, were best hypothesized as arising by enzymatic-induced chemical modifications occurring after the respective true amino acids were incorporated into proteins. Given this viewpoint, there was solid evidence for only 20 genetically specified amino acids.

We put this list of 20 amino acids to paper after the unexpected receipt of a zany communication from the celebrated, Russian-born, theoretical physicist George Gamow. While temporarily at the University of California, Berkeley, he was alerted to our second *Nature* paper by his physicist friend, Walter Alvarez. It was not, however, until Gamow had moved on to the University of Michigan that he found time to read it. Writing from the Michigan Union at Ann Arbor in July, Gamow correctly saw that the deep challenge ahead was how genetic information carried by the sequences of DNA bases was used to specify, say, that a cat grows up to be a cat and not a mouse. Most importantly, he saw that the language of DNA used as its letters the four bases A, G, C, and T. Intrigued that a number-theory approach might help solve how genes work, Gamow wanted to see Francis and me during his forth-coming mid-September visit to England. His hand-printed letter, how-

At the Institute for Theoretical Physics, Copenhagen, 1930. In the front row (from the left) are Niels Bohr, Werner Heisenberg, Wolfgang Pauli, George Gamow, and Lev Landau.

ever, had so many whimsical qualities that we did not know how serious he might be. As we both were to have left Cambridge by the time of his proposed visit, I filed his letter away unanswered, never suspecting that we would soon see him face to face.

By then we were looking forward to a meeting on proteins at Caltech that Linus Pauling (Sr.) had arranged for mid-September. I was going there anyway, but for Francis this invitation was a welcomed acceptance into the real world. Max Perutz, John Kendrew, his student Hugh Huxley as well as everyone important in protein crystallography would be there. Lawrence Bragg was invited in a special way. A letter came from Pauling asking him to be an Honorary Visiting Professor at Caltech and after the meeting to give some lectures on his work. When learning of this the chemist Jerry Donohoe, earlier of Linus's lab, Peter Pauling, and I, with John Kendrew's encouragement, saw a way for Francis's conversational skills to be used at Caltech. Towards this end we took assorted letters from Linus and assembled phrases to read as if written by him. We were especially pleased with the sentence, "Professor Corey and I want you [that is, Francis] to speak as much as possible during the meeting." The letter ended with the hope that Francis afterwards would

also lecture on his work as a Visiting Professor. Linus was soon coming
to Europe to go to a Chemical Congress in Sweden, where Peter was to
meet him. So Peter took with him a copy of the made-up letter that he
was to type on Congress stationery, affix to it a fake Linus signature, and
dispatch it back to Francis.

The following week Lawrence and Alice Bragg invited me to supper
at their new home out on Madingley Road. They were very pleased with
their new garden and before dinner, while walking me about, Sir
Lawrence was very apologetic about his behavior to Francis, saying it
was his worst mistake ever in misjudging great talent. After dinner over
coffee, Lady Bragg asked what "punk" meant. They said that young
Pauling had been there the week before, and after supper he sat down
and said, "I feel punk." I avoided a clear answer to Alice, suspecting that
Peter was lamenting his impending loss of Nina, the Perutzes' small,
pretty, blond au-pair girl soon to go back to Denmark. To say this now in
front of Bragg, however, would only compound Bragg's suspicion that
Peter's mind seldom turned to science. On several past occasions, Sir
Lawrence had gotten evasive answers from John Kendrew and Max
Perutz as to how young Pauling's first year of research was going. Both
knew that soon Peter had to get more serious or go home. But telling this
to Linus during his brief Cambridge visit to concede DNA defeat did
not quite seem right. So Bragg hoped that Peter would finally exude seri-
ousness upon coming to his home. But, later, all he could remember
from his and Alice's efforts would be "punk."

The next day, after telling Peter I had tried to put punk to rest, I
set out for the continent, going through Geneva and Zermatt on my
way to the International Congress of Genetics on Lake Como just
north of Milan. There, for the five-day meeting I was to be with my
co-conspirator on the *Nature* hoax, Boris Ephrussi, and his American
wife, Harriet. When I arrived at a big Bellagio Hotel, I found the Ephrussis talking to Baroness Edmund de Rothschild and her daughter,
Bathsheba, just back from the bacterial genetics course at Cold Spring
Harbor. Asking me to join them for supper, the Baroness later asked me
if I wanted some wine and I naturally said, "yes." Then I was asked to
choose it. So I picked out a 1912 bottle, not knowing much about it. A
white wine, it was orangish-yellow, being very old. She asked me how it

tasted, but I didn't know whether it was very good or very bad. Bathsheba was equally lost. So the Baroness, who almost never drank, took a taste and said, "Corky, send it back!" We repeated the charade a second time, and when the third bottle was opened, I said it was all right. They were all the same to me.

The next day I had lunch with Jeffries Wyman, then the scientific attaché of the American Embassy in Paris, whom I had met the year before outside Paris with the Paulings. Jeffries, a Bostonian, much preferred Europe over Boston, particularly after his first wife, a Cabot, died and a second marriage to an equally proper Forbes went quickly sour. We discussed what my future life in Pasadena might be like, and I said it would probably bore me. In contrast, he thought I might enjoy the more mannered Harvard, even though he never had fitted well into its Department of Biology, which was never quite first rank. Afterwards he wrote to his former colleague, George Wald, that I might be suitable for Harvard. As lunch went on I began to feel awfully sick, and soon was in bed with a fever of 104°F. For a brief while I feared I might be dying of polio picked up from the *spumone* I had eaten at the Como train station. But a day later I was out of bed, took the train to Zurich, and at the airport bought a watch for my mother. Then I flew to London and went up to Cambridge, apprehensive about the fate of our letter.

Francis and Odile Crick by then were already on the Atlantic on their way to Brooklyn. Earlier Francis had gotten our letter and was initially very pleased, writing immediately to Linus his acceptance. But there was a problem. Bragg had also been asked to speak after the meeting and how would they divide up the story? So Francis went up and saw the Professor. Bragg naturally was very depressed and went to see Perutz. As everyone in the lab knew about the letter, Max had to let on that it was a forgery. In response Bragg said, "Tell Crick!" Max, however, didn't have the heart to do it, and a week later Bragg said, "Send him up to me!" So Francis bounded up to the floor above where Bragg told him the truth. Francis rose well to the awkward occasion and immediately sent the fake letter to Pauling. Afterwards Peter wrote me saying that his father had recently deducted £5 from his allowance because of the letter to Crick. Then Linus wrote, "The letter caused me no end of trouble, because upon reading it I was convinced I had written it. But I recognized a grammatical error that I never make." Since then, every time I produce a split infinitive, I think of Linus.

None of the wives liked what we had done, and Elizabeth Kendrew curtly told me we had gone far beyond the bounds of decent behavior. At first we defensively said that they had not been subjected to one hundred renditions of the DNA lecture. The fun, however, was over and soon we felt guilty, and I worried what it would be like when I saw Francis. On my last afternoon in Cambridge, I had my final looks at Clare and the Chapel and Gibbs Building of King's. The long vac was over and by late afternoon the tourists were gone, and I was almost alone as I walked along the banks that I had first seen just two years before. The Wren Library of Trinity remained as overpowering as ever, and I felt somehow destined to return. Dining later by myself at the Arts Theatre, the next morning I took the train up to London to spend my last night in England at the Charing Cross Hotel before getting the boat train to Southampton.

After checking in, I almost instinctively began walking down the Strand toward King's. But bumping into Maurice Wilkins would at best be awkward, so I changed course towards Soho and then into Mayfair, where the fancier women of the night walked their leashed dogs to dispel any ambiguity as to whom they were there for. The nights then were no longer short and innocent English girls, still in their almost shapeless light frocks, looked cold as they tried to act as if the fragile English summer was still about. The evening was still early when I went back to the immense late-Victorian room that I was to have for the night. At least for the moment, my English life was over.

New Haven, Northern Indiana, and Pasadena: September 1953

MY RETURN TO the States in August 1953 was on S.S. *Georgic,* an old Cunard liner that in the summer was a cheap hotel for students on their way to and from Europe. The slow voyage took a week and happily provided much opportunity to drink bouillon on the decks with two "proper" Vassar College girls, fresh from their year away at Edinburgh University. I never let on exactly who I was, and I hinted only mildly that I had made a major discovery. It was as if we were Americans of the Henry James variety, to whom manners were the essence. As we approached New York, the girls increasingly turned their attentions towards three British students with the right-sounding names of Colin, Derek, and Malcolm, all headed toward studies at Ivy League schools. Obviously not yet in their social league, I considered whether instead of having RNA for my future, I should try to be a Republican and aim for the higher reaches of the Foreign Service.

The day before we were to dock, such thoughts stopped when I got a radio message from Avrion Mitchison telling me where I might meet him in New York. He was then in the States for two years on a Commonwealth Fellowship that followed his Oxford Zoology education. Originally we had met through his older brother, Murdoch, also a zoologist, who first introduced us at a Trinity feast in Cambridge. Av was interested in broadening his talents beyond the immunological orientation of his Ph.D. thesis and later took up my suggestion that he spend a year at Indiana University with the geneticist Tracy Sonneborn. But a

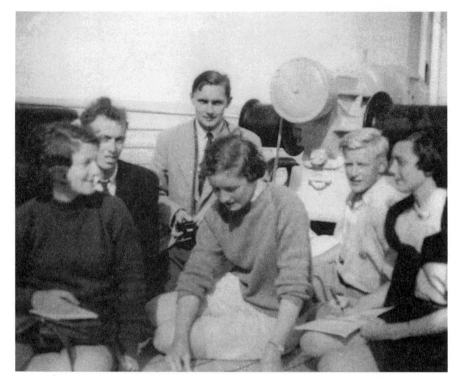

With friends on the deck of the Georgic, *August 1953: Lee Wakefield is
in the center, Margot Schutt is on the far right.*

year of Bloomington was enough, and from there he wrote me that, as in
Oxford, the nicest girls were married. Like me, he found the Hoosier
coeds good-looking but, at least to him, a bit inaccessible. So Av was
moving on to the Jackson Lab at Bar Harbor in Maine to resume work
on mice and the systems they provide for understanding immunology.

After my bags came off the boat and I gave the taxi driver my desti-
nation, he implied that it must be a whorehouse. Ten minutes later, I
was deposited at a seedy hotel just off Washington Square. With Av then
was Tony Richardson, soon to be the noted stage and movie director.
They had just driven through much of the States, going to every blue
movie and avant-garde play they could find. Av and his sister, Val, had
become friends of Tony in Oxford, from where he had moved to work at
the Lime Grove Television Studio in West London. This was to be the

only time we ever met. We disagreed about Henry James's *The Awkward Age* that he then had in his hands.

That afternoon Av and I left Tony, who was about to return to England, and drove in Av's beat-up car to New Haven to be with my father's physicist brother, Bill Watson, and his wife, Betty. Before my going to Europe, Betty, embarrassed by my bad dress and lack of tact, had quietly ejected me when I unexpectedly walked into a New York hotel suite where my cousin Ruth and fellow students from Smith College had gathered, all bound for Paris and Geneva. Now, three years later, many of my American words had turned into their English equivalents, and with DNA under my belt I was no longer regarded as a poor relation. When we talked after the visit, Av could not understand why I had been so hesitant in exposing him to Yale's way of life.

In Cambridge, Mass., after spending a night near Central Square, I nervously sought out Christa Mayr by going to her family's new home in part of a large early twentieth-century house near the Radcliffe Yard. But we found only her mother, Gretel, as Christa was with her father in Harvard Square. An hour later, we caught up with Ernst at the Cambridge Trust Company across from the Harvard Yard, and the conversation easily turned to the great eccentricity of Av's brilliant uncle, J. B. S. Haldane, whose fame as a geneticist mattered much more to Ernst than his long-time membership of the British Communist Party. Christa, to my distress, was off shopping in anticipation of Swarthmore College, where she was soon to start her first year. Soon after I was at South Station to board the train that was to take me back to the Midwest and my parents' home outside of Chicago.

When met by my mother, taking a day off from her job in the Admissions Office of the University of Chicago, I hid my frustration over not seeing Christa in Boston. Instead we talked about my sister Betty, who would be married in several days, and of the failing health of Nana, my maternal grandmother. She had recently failed mentally to the point that the local nursing home no longer wished to care for her. Later I took several long walks in the Indiana Dunes State Park, scrambling up the large mounds of sand that I had wandered over many times as a keen adolescent birdwatcher. Less than a mile away was the modest wooden home where my parents now lived after selling their small bungalow in Chicago's South Shore neighborhood. Mother and Dad had long wanted to live outside Chicago but had the means to do so only after my sister and I had finished at the University of Chicago. At last they were living among friends who shared their liking of the rolling green fields of northwestern Indiana.

After being home for almost a week, my parents drove me to Midway Airport for the long flight to Los Angeles. Under my arm was the red-and-green display model of the double helix, leading a stewardess to think I was an artist. So I felt pleasantly important until dirty yellow smog enveloped the plane as it descended over the mountains above San Bernardino to enter the Los Angeles basin. The acrid stench, which greeted me as I left the plane and rode through the palm tree–lined streets into central Los Angeles, grew even viler as the taxi ascended the Pasadena Freeway. In Pasadena, we turned off towards the Caltech campus, sited just north of the San Marino mansions through which much of the old wealth of LA had chosen to display itself.

In a few minutes we were at Caltech, my immediate destination being the Athenaeum, Caltech's imposing faculty club. I and many of the meeting participants, including Sir Lawrence and Lady Bragg, were to be housed there, and I entered knowing I would soon be among friends. Instantly I saw John Kendrew waiting for someone familiar to join him for supper in the tall, almost baronial dining room, dominated by a painting of Caltech's founders, including their Nobel Prize–winning physicist, Robert Millikan. John, now away from Cambridge for more than a month, was curious about how Francis had learned that the Pauling letter from Stockholm was a fake. Not knowing how Francis would greet us, we were relieved that the X-ray crystallographer David Harker came into the dining room with him. They had traveled together from Brooklyn, where David's lab had a million dollars to solve the structure of the small protein ribonuclease.

By then John and I were just finishing our dessert and soon excused ourselves for an evening walk through the carefully manicured Caltech campus, whose classical Spanish-styled buildings gave off the aura of ancient origins. In fact, its serene charm was the product of a frenetic post–World War I building program that had transformed Caltech from

Pasadena Protein Conference (September 21–25, 1953)

a provincial technical school into a world-class university for science in less than two decades. It was the mighty telescope, built upon nearby Mount Wilson to utilize the crystal-clear air of Los Angeles, that first gave Caltech a national visibility, and by capturing Robert Millikan from the University of Chicago to be its first president, it seemed natural for Albert Einstein to make several visits. Eminence in biology followed when the famed geneticist Thomas Hunt Morgan was recruited from Columbia just before the Depression took over. Even rich Californians worried then. But Caltech was far too good to be fiscally threatened for long, and after the end of World War II its upward momentum resumed. Although it had only 800 undergraduates then, it was in the same league as its much-older sister institution, the Massachusetts Institute of Technology, familiarly called MIT.

One of the major reasons for Caltech's ever-growing fame was Linus Pauling. A native of Oregon, and one of Caltech's first doctoral students, Linus used a Guggenheim Fellowship to go to Europe just as quantum mechanics came into existence. By then he was married to Ava Helen Miller, a fellow student from Oregon. Upon their return to Caltech, Linus quickly used the quantum thinking to revolutionize the world's idea on the nature of the chemical bond, publishing a seminal book under this name in 1939. Now, in effect, the Pauling family was Caltech's royalty. Much gossiped about and envied, few couples knew them intimately enough to know when they were expected to speak as opposed to listen. Linus and Ava Helen's inherent social hesitancy was unexpected, given the broad smile invariably displayed by Linus as he lectured about science to his fellow scientists or went into the outside world to dazzle the general public. With time, one realized that the Pauling charm was never expected to be returned in kind. And differences in opinion, no matter how lighthearted, were not easily taken.

First living modestly near Caltech, Linus's position since 1937 as head of the Chemistry Division gave them the means to move their four children to a large tetrahedral-angled, one-story home in the foothills. From there he drove to and from Caltech in an open Riley convertible that he had brought back from Oxford, where he was the Eastman Professor at Balliol in 1947–8. Linus had first become serious about biology in the mid-1930s when he began worrying about the nature of the chemical bonds holding antigens and their respective antibodies together. And just after the war, he was the first to realize that the disease of red blood cells called sickle-cell anemia had its origin in a molecular defect of hemoglobin, the protein in red blood cells used to transport oxygen through blood vessels. But it was his 1951 proposal of the α-helix as the basic fold for the polypeptide chain that was to give Linus his greatest impact on biology. The meeting that we were about to attend had been organized in part to let Linus again display its deep beauty. I'm also sure that for several months Linus looked forward to showing off his new model for DNA. But now, to the obvious pain of Ava Helen, the DNA story was for Francis and me to relate.

At the meeting, Linus made a big point of DNA's guanine–cytosine base pair being held together by three hydrogen bonds, one more than Francis and I proposed in our original *Nature* paper. Then, not knowing the exact guanine structure, we thought the third bond might be much weaker than the other two and so left it out. Later experiments that demonstrated the high thermal stability of GC-rich DNA samples proved that Linus's chemical intuition was again right on course.

Linus Pauling (left) and George Beadle at Caltech, late 1953

All in all, this was a useful scientific gathering, even though most of the talks merely rehashed familiar facts and concepts. The participant most genuinely excited about what the next few months might bring was Max Perutz. Over the summer he had learnt of a Dutch result that might let him crack the phase problem, which was then preventing him and John Kendrew from solving, respectively, the hemoglobin and myoglobin structures. To my relief, Francis seemed not that put off by our practical joke, making sensible remarks on numerous occasions. On the other hand, he was clearly off the manic high that had been such a characteristic of his life in the previous several months.

He and Max Delbrück argued the first evening whether the two strands of the double helix could untwiddle fast enough to let them separate. Soon a $5 bet was made with the Caltech chemist Vernon Schoemaker to hold the money in escrow until the matter was settled. Francis and I then saw no way to distort the individual helically folded DNA chains so as to let them lie side by side as Max wanted. But until there was a rigorous X-ray proof for the double helix, Max was not likely to concede and the money seemed destined to remain in Vernon's pocket for a long time.

Late the second afternoon, Linus and Ava Helen held a garden party in their home at the end of Sierra Madre Boulevard. Unfortunately it was a week too late for their daughter, Linda, to be there. Now aged 20, and 18 months younger than her brother Peter, she was back at Reed College starting her senior year. Four years before, when I first came to Caltech for summer research, Linda was at a late-night party given in a foothills house then shared by several postdocs at Caltech. Her glamorous blond personality stood out and, to my disappointment, she disappeared before I could get close. Later Peter told me that Linus and Ava Helen wanted Linda to grow up unsaddled by the conventional middle-class beliefs that kept most children from developing into interesting personalities.

With Linda away, I did not have high expectations for this Pauling party. There were almost no women associated with Caltech and most meeting attendees were there without wives. By then I had already met Pauling's postdoc, Alex Rich, three years my senior, who came to the party with his wife, Jane. Jane had been raised in Cambridge, Mass., where Alex had met her when he was at medical school at Harvard. Seven generations of Jane's family, the Kings, had gone to Harvard, and she had no trouble in sensing that I must be going through culture shock in giving up the real Cambridge for a southern California where money matters more than words or ideas.

Because the smog was still ferocious, we could not enjoy the view of the mountains above us and, to pass the time, Jane and I started telling John Kendrew how American politics operated. After the wife of Lee Dubridge, the physicist president of Caltech, joined us, I exaggeratedly said that American politicians were all corrupt and lousy—a natural remark for someone who grew up in Mayor Kelly's Chicago. She looked pained and said, "But don't you respect President Eisenhower?" Instantly I replied, "No." Little did I know that Lee Dubridge had been part of the delegation of senior Republicans who had gone the year before to Eisenhower asking him to challenge Ohio's Senator Taft for the Republican nomination as President. Unclear then was which of us was the more gauche, she for the question or I for the answer.

Pasadena, Northern Indiana, and the East Coast: October 1953–January 1954

WHEN THE PASADENA meeting on protein structure finished at the end of September, the full horror of being in Pasadena hit me. Not knowing how to drive a car, much less owning one, I was effectively confined to the girlless Caltech campus and had to continue living at the faculty club, the Athenaeum. Although the bacon and eggs were faultless, the *Los Angeles Times* was not the best way to start a smog-filled day, with its pages seeing hidden communists behind the thoughts of any liberal Democrat. Each day, when breakfast ended, I became part of Max Delbrück's ground-floor phage group in the 1930s-style Kerckhoff Biology Building. Two years previously I would have thought this to be the best of all possible worlds. Then phages were it and I would have died for them. But now I had a desk in a room assigned to several of Max's graduate students, both working on, at best, dull topics. In particular, Gordon Sato was doing kinetic experiments to learn more about how the amino acid tryptophane helps phage T4 attach to *E. coli*. There was no way to imagine this problem, even when solved, as ever being more than a further addition to academic trivia.

With my new interests, I belonged more appropriately to Pauling's lab in next-door Crellin Laboratory. But I knew it would be unwise to have my fate determined by Linus. With my long-term goals unabashedly that of a biologist, I wanted to be judged by those with common objectives. Happily I saw a way of pursuing RNA's structure while remaining independent of Pauling. By collaborating with Alex

Rich, who worked in his lab, I could stay officially under Delbrück and not need Pauling's blessing. There was no difficulty in persuading Alex to turn to RNA because there was no longer any reason for him to take X-ray photos of DNA, a task he had begun after Linus proposed his triple-helical structure. Instead, RNA was wide open and an obvious long-term objective for the new laboratory he would be starting a year hence at the National Institutes of Health (NIH) in Bethesda, Maryland. Soon I was writing RNA chemists to send us their best RNA preparations for X-ray structure analysis, but until they came, I wanted to maintain the facade that I was still excited by phage work.

At this point, my mind became increasingly dominated by the fear that I would soon be in the U.S. Army. Although I had been occupationally deferred all through the Korean War, which had just ended, my draft board in a South Chicago steel-mill district had decided that my time had come and made me 1A. I first heard of my reclassification while temporarily home with my parents but waited to respond until I arrived at Caltech. Immediately George Beadle, the geneticist head of the Biology Division, wrote to the draft board asking that my occupational deferment be restored. His plea, however, went nowhere, and I faced the possibility of soon starting two years of military service without being able to do one push-up. So I was easily persuaded by Alex Rich to apply to be an officer in the Public Health Service, assigned to work with him at NIH. Quickly I filled out the appropriate application forms and went into LA to take the necessary physical exam. Afterwards, I was uncertain whether I should have been happy at passing it. If they had rejected me for my flat feet, the army might later do the same.

The smog occasionally vanished, and I could then appreciate why the pre-automobile Pasadena had been so attractive to retirees born in the Midwest. But even the occasional tennis match on the Athenaeum courts only temporarily made life seem worth living. While my hosts in the Biology Division went out of their way to try to make me feel at home, I kept longing for unexpected responses to mealtime remarks. Their absence would have mattered much less if I had found a California girlfriend. But most graduate students seemed to be married and there was no obvious girl-seeking group I could be part of. So my thoughts often turned to Christa Mayr, to whom I had written the moment Pauling's protein meeting ended. She soon replied, but her glowing descriptions of Swarthmore life made me feel uneasy.

After several emotionally empty weeks, the mail finally brought an RNA sample for Alex and me to put before the X-ray beam. By using a needle, I could easily draw out long, thin, birefringent fibers after adding a little water to the powder-like pure RNA. With luck, these fibers would contain long, thin RNA molecules packed regularly next to each other. I eagerly passed on my first such fiber to Alex to take into the X-ray room. As he and Jane never rose before noon, only late the following afternoon were we able to examine the resulting X-ray film. Depressingly, it contained a blurred diffraction pattern no better than the one I had obtained six months before in Cambridge using plant viral RNA that Roy Markham had given me. At dinner with Jane, Alex and I reassured ourselves that better RNA samples might soon materialize and let us begin serious model-building. But as I walked back towards the Athenaeum, my thoughts could focus only on the return of the smog, my 1A draft status, and the fact that there was only one secretary to stare at when drinking coffee at The Greasy Spoon—the coffee shop on the Caltech campus, where, twice a day, I used to go to pass the time.

At this point in my Caltech life, I welcomed the arrival of Bob DeMars, who after finishing his Ph.D. with Salvador Luria had come West to be a postdoc with Max. Now temporarily living in the Athenaeum, he needed to acquire a car but was short of cash. So he accepted my offer to buy jointly the '41 Packard sedan that Manny's parents no longer used, possibly because it emitted an awful stench. The asking price of $125 seemed excessive, but Max said take it or leave it and we saw no point in arguing. Our first trip was a Sunday drive along the Angeles Crest Highway up to Mount Wilson, where, after our arrival, I decided it had to be less scary driving than looking over the sides of steep drop-offs. Later, in Bob's presence, I began practice driving on the curved, deserted streets of San Marino. But I never came close to passing my first driving test, totally panicking after aborting a reverse parking maneuver.

I was still without a license when I spotted a tallish girl with possible real flair at The Greasy Spoon. She came only irregularly, and it took me more than a week to discover that she was a research assistant at the

Phytotron, Caltech's environmentally controlled, new super green-house. Upon finally meeting her, I learned that her family was from the East and that she had gone to Vassar College. Seemingly now unat-tached, Rachel Morgan's demeanor reflected old money—she might be the way to bring back the mannered days of England. So we used her car to go to dinner after I gave a Biology Division seminar on the double helix. That was my first occasion to wear a new checkered English suit that I picked out for myself just before I left Cambridge. We went to The Stuffed Shirt—Pasadena's red-leathered, English-type restaurant off Green Street. There I learned that although her grandfather had been very rich, he was not J. P. but E. D. Morgan, whose large estate was also on Long Island. As the meal ended, I tried to set up a new date, but she controlled the occasion and I was left dangling. Then, as she dropped me off at the Athenaeum, her mood changed and she said that she wanted to cook for me in the near future.

My new sky-high morale was tempered the next day when my Swiss co-hoax perpetuator, Jean Weigle, then in his Pasadena winter phase, warned me I should not count on Rachel's affection. She had recently been very close to a just-departed South African botanist. In any case, I no longer wanted so desperately to leave Pasadena and was nervously apprehensive when Bob DeMars drove me to a Los Angeles Armory for a pre-induction physical. My hopes went up when I failed the exam on machine tools and did poorly on recognizing upturned shapes. They zoomed even higher when I was ascertained to be the most emaciated of a group of 35 nude males. I was taken away to a cubbyhole used by a Beverly Hills psychiatrist. At first sight highly civilized, he questioned me as to how I would enjoy army life and then to my reaction to pretty girls. Then he wrote some words on my form and went on to his next assignment, a very fat Mexican-American. Out of his sight, thinking he was on my side, I glanced at the psychiatrist's verdict: "The aesthetic type—capable of military service."

This, however, was not an immediate death sentence because by then Caltech had found it could appeal at the California level and, if necessary, request further review by a presidential panel. So I was safe until late spring, by which time a Public Health Service Commission should be my safety net. So I relaxed over the Thanksgiving holiday, camping with the Delbrücks among the Joshua trees in the high deserts east of Palm Springs. Afterwards, I had only a few days back in Pasadena before Bob DeMars and I drove up to the University of California at Berkeley. Gunther Stent, since 1952 its phage hot shot in the new virus lab there, had arranged that I be paid 40 large dollars to give a Friday afternoon Biochemistry Department lecture on the double helix. Located near the top of the campus, the Virus Laboratory had a marvelous view towards the Golden Gate Bridge and I could see that California life need not mean Pasadena smog. Moreover, there were good-looking girls everywhere, even among the overflowing audience at my talk.

I turned from jocular to serious when Gunther brought me into the office of his department chairman for the usual 10–15-minute interval that Wendell Stanley allotted to speakers before his department's seminar. Quickly I gushed forth on the beauty of his building, trying to erase the bad impression I created two years before in Copenhagen. After Stanley's lecture on tobacco mosaic virus (TMV) at the Polio Congress, I went with him among others to dinner at the home of physicist Niels Bohr. I still held in my mind the pretty electron microscope picture of TMV that had featured in Stanley's talk that afternoon. To make small talk, I remarked that this structure was almost as beautiful as the smashing new Rome train terminal, which I had passed through when living in Naples. To my dismay, Stanley became very defensive under the false impression that I was comparing his new Berkeley Virus Laboratory, not his TMV, to a train station. Later, among his lab scientists, I tried to exude low-keyed maturity, knowing that if the army draft vanished I might welcome a job offer from Berkeley.

On the way back to Pasadena, Bob and I stopped at the last moment at Stanford, allowing me to give an informal seminar before the geneticists associated with Ed Tatum. When we reached the Los Angeles region, it no longer seemed so intolerable, especially in view of the forthcoming supper at Rachel Morgan's apartment several blocks from Caltech. That evening, when shown a photograph of her family in front of her grandparents' Long Island mansion, I was so beguiled by her tallish blond presence that I momentarily stopped thinking about the draft. Soon after, we went into Los Angeles to see the Sadler's Wells Ballet perform. But the huge Shriner's Auditorium was not Covent Garden, and that evening's ballet, *Silvia,* proved too slight to help make the evening click. Afterwards, I was also put off by Rachel's matter-of-fact tone as we drove back to Pasadena.

I was soon diverted from the thought that I might still be girlfriend-less in southern California by the prospect of being in Chicago for Mother's birthday on Christmas Eve. That evening at home we had oyster stew, a family custom that dated back to my mother's early years. After a week in Chicago, I took the train to Boston for the annual meeting of the American Association for the Advancement of Science in the Old Mechanics Hall. Francis Crick was coming up from Brooklyn while Av Mitchison was driving down from Bar Harbor. The three of us stayed two nights in Cambridge in Hugh Huxley's second-floor flat on Harvard Street near where Trowbridge Street crosses it. Over the dinner he cooked the first night, Hugh talked about the thin-section electron micrographs that he was taking in F. O. Schmidt's MIT lab. He had been there a year as a Commonwealth Fellow following his Ph.D. work at the Cavendish lab in Cambridge, England. Hugh then hoped to shake up muscle biology by showing at the molecular level how muscles contract.

The next morning, I went over to the Mayrs' house to learn that they were purchasing a 200-acre farm in New Hampshire and would start spending their summers there instead of at Cold Spring Harbor. Later, Christa came with me to the Mechanics Hall, where she saw Francis in action for the first time, talking about his meeting in New York with George Gamow, whom he found not the nut cake we guessed from his letter of the past summer. Then, most unexpectedly, Christa and I found ourselves face to face with Paul Weiss, who had taught me invertebrate zoology at the University of Chicago. I had last seen Paul in New York when I was interviewed for the fellowship that enabled me to go to Copenhagen to learn biochemistry. But 18 months later, during my last days in Copenhagen, he had angrily disapproved of my request to transfer to Cambridge, asserting that I was unqualified to take up X-ray crystallography. Now I had the last laugh, and enjoyed seeing Weiss's inability to be even hypocritically pleasant as he beat a hasty retreat. Christa that day was great fun to be with. Again her face and voice made butterflies rumble through my stomach, but we parted as friends, not lovers, when we said good-bye at her parents' home.

The following afternoon, Hugh Huxley drove me out to the home of Lee Wakefield, the more straightforwardly nice of the two Vassar girls I'd met on the boat back from England. Then I learned that she was even more of a "proper Bostonian" than I had guessed on the *Georgic*. Her mother was a Forbes of the branch that summers off Woods Hole on Naushon Island. From Lee I learned that neither she nor Margot Schutt had yet planned what they would do after graduation. On the boat, Margot appealed to me the most, possibly because she kept hinting at her need to fly away from the confines of good taste. But she never answered the letter I posted to her at Vassar soon after I got to Caltech.

On New Year's Day, armed with a box of chocolates, I rode the New Haven train down to Yale to spend the night with my aunt and uncle. Aunt Betty, seeing that my hair had been recently cut and correctly judging that I was no longer likely to embarrass them, took me to the Lawn Club, cheerfully informing everybody within hearing distance that her nephew Jimmy would soon be an important professor somewhere. The next morning I continued on to New York for a night with the Cricks. They were living on Fort Hamilton Parkway in a dreary 1920s apartment about as far away from Manhattan as you can get in Brooklyn. They were not even close to where Francis was working at the Brooklyn Polytechnical Institute. Despite the million dollars behind David Harker's lab, its ribonuclease project had gone nowhere, and Francis was there to give it real brain power. Odile, then several months pregnant, and I initially discussed what name she and Francis should give their second child. If it was a boy, I offered the English strength of Sebastian Trumpington Compton Crick. But if they had a girl, I was keen on Adenine Crick with the thought that she could be called *Addy*. We later went over the virtues of American life—doughnuts, frozen orange juice, and such like. Words, however, could not transform the Cricks' daily life into an experience worth crossing the Atlantic. I also had the feeling that Odile had not forgotten the faux Pauling letter.

Somewhat subdued, I took the hour-long subway ride back to Manhattan to take the train down to Washington for my first glance of George Gamow. Before meeting him, I was to visit the NIH in nearby

Bethesda to see whether my commission in Public Health would soon materialize. Talking with their top brass was likely to be a heavy occasion because I would have to give the impression that I looked forward to being part of the Public Health Service, not that I was trying to avoid the draft. So ending my visit with zany Gamow paradoxically seemed the way to stay sane.

Bethesda, Oak Ridge National Laboratory, and Pasadena: January—February 1954

THE NATIONAL INSTITUTES of Health (NIH) in Bethesda, Maryland, was not the big prison I anticipated. Its bosses were keen to get me, no matter for what reason, and I took comfort that so many of its scientists worked on problems with no immediate clinical consequences. Nonetheless its new Building 10 felt like a giant hospital, with its several hundred beds and shape of a giant, brick ocean liner. Happily NIH didn't want me to appear until mid-summer even if my commission came through, as they expected, by late winter. My two nights in Bethesda were spent at the home of the ion-channel physiologist, Adrian Hogben, whom I had known when we were both postdocs in Copenhagen. From him I got a feeling that NIH, despite not feeling like a university campus, had real intellectual expectations.

The next morning Adrian left me off at the Watergate Inn where Rock Creek empties into the Potomac River. There George Gamow instantly recognized me. From Francis Crick he knew more what to expect of my frame than I did of his. His high-pitched voice did not go with his generous bulk. Now 50 years old, the very tall thinness of his Russian youth had given way to a middle-age girth, accentuated by more alcohol than generally compatible with high-powered manipulation of mathematical symbols. Instantly I sensed his excitement with his new life as a biologist. Not that his physics was going badly. In fact, he was leading the world in cosmological insights. Five years before, he and his graduate student, Ralph Alpher, had calculated how the chemical elements were built up by neutron capture following the Big Bang when the Universe formed. To announce their answer, they prepared a short note to *The Physical Review* entitled "The origin of chemical elements." Gamow was intrigued by the possibility of the paper being the first authored by alpha, beta, and gamma, provided that he could convince his sometime fellow cosmological explorer, the great physicist Hans Bethe, to add his name to the manuscript. Consent came swiftly from Bethe, not ordinarily given to Gamow-like foolery but obviously bitten by the possibility of being associated with a seminal contribution to the first several minutes of the Universe. Initially the paper's significance was not widely appreciated. The fact that it was published on April 1, 1948, confused even those who knew Gamow's past tricks.

As Gamow and I sat down, he told me that everyone called him Geo ("Joe"), and he pressed upon me a gift, the Japanese translation of his latest book, *Mr. Tompkins Learns the Facts of Life*. Telling me it was inscribed, I opened the front cover only to find the inscription, "I've fooled you—open the other side." Geo delighted in his joke's success, guessing correctly that I had no reason to know that Japanese books start where English books end. Then, being sure that I had a Scotch and soda, he started talking about a set of rules that would allow the 4-letter (A, G, T, C) DNA alphabet to be translated into the 20-letter (amino acid) alphabet of proteins.

Under this scheme, proteins were assembled from the 20 different amino acids directly on the surfaces of DNA molecules. Immediately I knew there was no chance of this being right because no DNA is found in the parts of the cells where proteins are assembled. Instead, the surfaces on which the amino acids are linked to each other have to be RNA molecules. But my protestations on this point fell on closed ears. Geo wanted RNA and DNA to have the same structure, even though I told him that RNA gave a different and characteristic X-ray diffraction pattern. Already he had sent a note to *Nature* stating that a precise genetic code exists in which every amino acid is specified by some set of the four bases (A, T, G, C) found in DNA. Because there are 20 amino acids, some if not all had to be specified by more than two bases because the number of combinations of two bases is only (4×4). If, however, each amino acid was determined by triplets of bases, there would be many more combinations ($4 \times 4 \times 4 = 64$) than specific amino acids. Already Geo had a trick for reducing the 64 sets into 20 groups, but because I thought all DNA-based schemes must be bunk, I let him talk without really listening.

By the time Geo drove me to the airport, I was in a mild alcoholic daze even though early in the lunch I had stopped trying to match with Scotch and water Geo's rapid downings of neat whisky. I was on my way to Knoxville, because Alex Hollander in Boston had offered me the princely sum of $150 to stop by the Oak Ridge National Laboratory in Tennessee to talk about the double helix. There Alex was running the U.S.'s biggest effort to measure the genetic consequences of ionizing radiation. My flight there was especially fun because a very pretty stewardess became intrigued by my copy of *Satan in the Suburbs,* Bertrand Russell's new book. At Oak Ridge, security passes were needed to move in and out of the giant factory-like home of Alex's empire. Once inside, however, the atmosphere was academic, and it was fun to begin thinking about genetic damage in terms of a real molecule, DNA. By then, however, I was travel weary, and the final flight back to Los Angeles passed slowly, even though John Wayne and his Peruvian girlfriend, Pila, on their way back to the Pacific Coast, joined the plane at Dallas.

Back in Pasadena, I again had more time than I knew how to fill. Happily my morale went up when I got my driver's license and could buy my first car. In a Colorado Avenue used-car showroom, I found a three-year-old white Chevrolet convertible that I soon parked with semi-pride next to the Athenaeum. But I was envious when I saw next to it the Jaguar of my fellow resident, William Shockley. He was at Caltech briefly between his departure, from the Bell Labs and from his wife, and his move to the Bay Region to start up a company to exploit the transistor, which he had helped invent several years before. Conversationally Shockley did little to enliven the tedium of Athenaeum evenings, and I could only look forward to evenings when I was invited out to dinner. Most frequently it was to the Riches', who were living in a comfortable old house off Oak Grove. Jane had little interest in subtle food but always could be counted on to listen to my complaints about single life in a place (Pasadena) that had the highest concentration of women over 60 than any other American city. All the seemingly suitable girls she knew were 3000 miles away, but their existence gave me hope there might be life after Pasadena.

By then I knew I would go nowhere with Rachel Morgan—the girl I met at The Greasy Spoon soon after I arrived at Caltech. Soon after my return from the East Coast, she told me I could never be important in her life. This news kept me from sleeping for several nights, after which I turned to Alex Rich for possible help in getting sleeping pills. But he warned me that they could prove addictive and that I should avoid them. Instead, I took comfort in hearing by mail from Peter Pauling that his parallel search for the perfect girl in Cambridge was also going nowhere. At the summer's end, he got unmistakable hints from his parents that his life would improve if he found a nice girl and got married. In fact, Linus dangled in front of Peter the inducement of a higher allowance. In contrast, Linus's friend the British zoologist Lord (Victor) Rothschild, went no further than asking Peter, "How was his sex life?" To which Peter replied, "Non-existent."

Now Peter wanted to buy the radio I had left with him as well as my winter coat with my copy of Kinsey's 1953 report, *Sexual Behavior of the Human Female,* thrown in for nothing. Over Christmas, he had been on a semi-lonely voyage to Greece that had had some awkward moments. He had just vowed to raise the minimum age to 21, not knowing whether this rule would accommodate the pretty and fresh new technician in the Cavendish lab who called everybody, including Max Perutz, by their first names. And Peter was ambiguous about what to do if his sister Linda came to England—that is, if she didn't marry the college friend who was keeping his parents in the dark about their intentions.

With the winter rains falling much of the time, Caltech at last seemed an appropriate place for inside work. For much of January, Alex and I were encouraged by our work on RNA. The RNA samples that had recently arrived gave better X-ray diffraction patterns than he'd seen during the fall. But our elation proved temporary when we later could not further improve our results. Although the RNA X-ray pattern was indeed specific, and not obviously related to DNA's, it remained too fuzzy to give us hope that a well-defined RNA structure existed. What was worse, the same pattern emerged from RNA in which the base ratios almost smacked of complementarity (A = U, G = C) as well

Outside the Riches' Pasadena house, 1954: (from left to right) Jack Dunitz,
Giovanni Giacometti, JDW, and Alex and Jane Rich

as from RNA where the base-pairing rules were clearly violated. On most days I was convinced that RNA must be single-stranded, arguing that the samples with base compositions hinting at double helices reflected the fact those RNAs arose by base-pair-mediated RNA duplication in the cytoplasm. My argument was admittedly far out and never convinced Alex.

Max and Manny Delbrück tried to rescue me by arranging a camping trip to the desert with Doriot Anthony, a young flutist from Boston, temporarily studying in LA and who frequently came to the Delbrücks' ranch-style home on Oakdale Avenue. But she proved as uninterested in me as I was with her. More girl hope came after George Beadle told me that Gary Cooper's daughter, Maria, was thinking about being a biochemist and was coming with her parents to see Caltech the following week. The "royal visit" occurred just after lunch, and most of the Biology graduate students and postdocs eagerly awaited their walk down the halls of Kerckhoff. Suddenly they arrived with George happily in charge of a trio not up to our *High Noon* expectations. Gary was much less impressive than his well-dressed wife, while his daughter's still-slender form was not that of Grace Kelly. By the time they left, I feared this might be the only Hollywood moment Caltech would bring into my life.

About this time (February 1954), Geo Gamow wrote me from the train taking him across the western states for his spring residence at the Berkeley Physics Department. He wanted to know more about RNA and was rightly confused as to its role in certain viruses where conceivably it, not DNA, was the genetic component. After he picked up his new car, he planned to drive to Pasadena, to see the Delbrücks. Max had known him since their early 1930 Copenhagen days when they were together in Niels Bohr's Institute of Theoretical Physics. It was hard for Max not to be jealous of his high-spirited Russian companion, who, after growing up through the Revolution in Odessa, studied physics in Leningrad (now St. Petersburg). By the time Geo reached Copenhagen, he was already famous for his tunneling theory, which used quantum mechanics to explain unstable atomic nuclei. Then he was treated as a hero in his home country, with one Russian paper writing that "a son of the working class has explained the tiniest piece of machinery in the world." Although *Pravda* earlier had honored him with a poem, upon his return to Russia in 1931, he found himself in an intolerable situation. Quantum mechanics was being denounced as an enemy of dialectical materialism, as was Einstein's theory of relativity.

By then married, Geo concentrated on ways of returning to the West and made abortive moves to escape, first across the Black Sea and then from Murmansk into Norway. Thwarted in both attempts, freedom came to him through an invitation to a Solvay Conference in Belgium that he managed to have extended to his wife, Rho, who had once been a physicist. By not returning to Russia, the Gamows caused an uproar, and Stalin ordered that no more Russians be allowed to attend international scientific meetings. At the Solvay Conference, both his French and non-French speaking colleagues demanded that he cease trying to speak French. Geo badly wanted to move to the States. Within a year of his escape from Russia, however, an opportunity came when he received an invitation to join George Washington University in Washington, D.C. So Geo bought a ticket to Seattle before discovering that he was to live on the East Coast.

Everywhere he insisted on having fun, even though many colleagues resented his having a good time at the expense of good taste. While in Copenhagen, his practical jokes culminated in an assault on the editor of *Naturwissenschaften,* the German equivalent of *Nature.* Upon reading a dreary, though legitimate, paper from India, George persuaded several European physicists to write to its editor that *Naturwissenschaften* was the victim of a hoax and had published purposeful trash. Niels Bohr, in particular, was not amused by this Gamow foolery and made him go to Berlin to apologize in person.

Geo now got much enjoyment from writing books for popular audiences, which he illustrated himself and filled with silly rhymes. Central to their contents were the adventures of Mr. Tompkins, whose initials were C (the speed of light), G (Newton's constant of gravitation), and H (Planck's constant). A bank teller, Mr. Tompkins would fall asleep during his physicist father-in-law's lectures and in his dreams encounter the objects of these lectures. Relativity, for example, was experienced while driving in a car at the speed of light of only several miles an hour. Unfortunately, Geo decided to tackle biology before the double helix was born, and his latest book on the facts of life that I had just read in English was not the success he hoped for. While he moved successfully among genes and chromosomes, Mr. Tompkins got mired in the red cells of our blood vessels and his dreams never reached beyond ideas that most educated people were already very familiar with. Unlike his prior popular books that sold well to students as well as adults, Geo's biological effort bombed and was never reprinted by Cambridge University Press.

Just before taking the long train trip west, Geo had written up the details of the genetic code that he had hinted about in his preliminary note to *Nature.* He'd been elected to the National Academy of Sciences (NAS), and he wanted to submit his first biology manuscript to its journal, *The Proceedings of the National Academy of Sciences of the USA,* or *PNAS* for short. Titled "Possible mathematical relation between deoxyribonucleic acid and proteins," Geo added as co-author his longtime companion, Mr. Tompkins. So when Merle Tuve, the geophysicist editor of *PNAS,* received Gamow's manuscript, he assumed that Geo's newest fun was at the expense of the Academy. Because Geo lived close by, Tuve asked him to drop by his office at the Carnegie Institution of Washington's Department of Terrestrial Magnetism. Implying that unnamed biologists were offended, Tuve warned Geo that although as an Academy member he could submit anything as long as it concerned the physical sciences, biology was out of bounds. The manuscript was returned to Geo, who put it in a new envelope and sent it for publication to the Royal Danish Academy, to which he recently had also been elected a member. Wanting publication more than fun, he removed Mr. Tompkins's name. Later he sent reprints of it to all the biology members of the NAS writing on each, "Best regards from Geo." Only then did his friend Harold Urey, the discoverer of deuterium, learn of this censorship and threaten to open up the affair. By then, however, Geo had other ways to amuse himself. If the Academy felt the need to look as well as be serious, that was their business. His role was to have a good time no matter the consequences to the ethos of science.

Pasadena: February 1954

SOON AFTER GEO arrived at Caltech in February 1954, I brought him to Max Delbrück's house where Vicky Weisskoff and I had also been invited. Vicky, now a professor at MIT, Max, and Geo had long known each other as theoreticians, and their dinner conversation initially centered on the Swiss-domiciled Wolfgang Pauli, whose legendary rudeness rivaled his formidable intelligence. Slowly Geo began to dominate the occasion by revealing his new manuscript with Mr. Tompkins's name still attached. Vicky and Max tried initially to follow his argument but soon got lost. Max announced that Geo would give a seminar the next afternoon on his genetic code and the intervening hours would give us time to look over his paper and prepare to argue where we lost his reasoning.

Soon, a number of Max's Caltech friends and colleagues arrived for coffee and dessert, giving Geo the opportunity to move on to his limericks and card tricks that easily fooled most of us through his deft sleight-of-hand maneuvers. Only Vicky did not join in the fun. He had seen Geo pull off these same tricks before physicists on more occasions than he wanted to remember. Hearing Geo's high-pitched voice squeal with delight was not Vicky's way to end a civilized evening. The next afternoon, Vicky's intellectual curiosity dominated his emotions, and he joined us in the small lecture room across from Max's first-floor office. Soon we realized that Geo's scheme would be proved or disproved over the next year by newly appearing amino acid sequences.

Gamow's brainchild was an overlapping code in which a given base pair was used to specify more than one amino acid. This was not a wild idea because the repeat distance (3.6 angstroms long) of amino acids along extended polypeptide chains was so similar to the repeat distance (3.4 angstroms) of base pairs along the double helix. Conceivably polypeptide chains contained the same number of amino acids as base pairs in their DNA templates. But because at least three base pairs must be needed to code for many, if not all amino acids, the amino acid sequences in proteins could not be a random collection. Instead, each amino acid would have only a restricted number of amino acids as their immediate neighbors. Up until now, however, too few proteins had had their polypeptide chains sequenced to reveal whether certain amino acid arrangements really were forbidden.

The next day Gamow was off to the Rand Corporation, the think-tank in Santa Monica designed to advise the Pentagon on the techno-logical future of war. Geo still maintained close contact with the world of top-secret weaponry that he had first come into contact with at Los Alamos during its push to make, first, atomic and then hydrogen bombs and was a long-time close friend with Hungarian-born physicist Edward Teller. Talking to generals and their minions was fun for Geo as well as a source of a third of his income, with an equal amount coming from his books. Thus, although his salary at George Washington University was never commensurate with his intellectual stature, he could have an exis-tence in which the pursuit of intellectual excitement was his primary objective.

On Sunday, Geo and I sought out a beach on which to sit and hope-fully meet some pretty faces. He drove us in his new white Mercury convertible that he called Leda, and he chose Long Beach as our desti-nation, expecting long stretches of white sand. But, on arriving there, its only inhabitants, to our dismay, were the oil derricks that dominated both the beach and the neighboring waters as far as we could see. Geo tried to cheer me up by saying that I reminded him of his Russian friend of 20 years before, the already fabled theoretician Lev Landau and took a photo of me to mark the occasion.

After he had left for Berkeley the next day, I was happily realizing that my life need not be that of a Chekhov hero exiled to a provincial town

" Feeling like a little boy collecting peables on te ocen shove " (Lord Kelvin).

JDW photographed by George Gamow on Long Beach, California, February 1954

devoid of style and culture. To start with, the winter rains banished the smog on most days, letting me enjoy the San Gabriel Mountains, which rise up massively to the north of Pasadena. And a growing coterie of intellectuals, attracted to Caltech by Max Delbrück and Linus Pauling, gave subsequent camping trips to the Joshua Tree a lively air. Among these new intellects were several physicists working on the response of Max's mold *Phycomyces* to light. Why Max had now moved on to light responses mystified those of us excited only about genes and how they function. Max's mere interest in a problem, however, lent it a cachet among physicists more at home seeking the truth mathematically from the shapes of dose-response curves. In contrast, the chemists round Pauling were there to deal with well-defined molecules.

Thus a newly arrived Oxford theoretical chemist, Leslie Orgel, quickly joined forces with Alex Rich and me in focusing on RNA. As a prize fellow at Magdalen, Leslie had a major intellectual success analyzing newly discovered organic molecules that trapped iron within cage-like structures. Although there obviously would be other new classes of molecules to find and understand, Leslie let us know that the prior work of Pauling and his arch rival, Robert Mulliken of the University of Chicago, may have left very little new theoretical cream to be skimmed off. Going on as a pure chemist was likely to involve more hard work than was justified by the small gems yet to be discovered. So moving on to molecular biology and macromolecules like RNA made more sense than continuing bread-and-butter theoretical calculations. Leslie was at our side as Alex and I spent much of February 1954 going through more and more RNA samples, one of which we hoped would yield an interpretable X-ray pattern. No real improvements came over our earlier January tries, however, and there seemed no good reason not to publish our results, inherently unsatisfying as they were. So our short manuscript describing the essential features of the RNA X-ray pattern was dispatched to Max Perutz for him to pass on to *Nature*.

Happily, there was now Richard Feynman to share our frustrations. As the cleverest of Caltech's minds, Dick, then just 35, never hid how hard it was for him to do innovative physics. Total relaxation between his bursts of intellectual activity was a necessity. The fame of his bongo-drum days preceded our first meeting, and he had recently remarried after a long single period after the immediate post-war death from tuberculosis in Los Alamos of his young, Brooklyn-born wife. Rumor had it that his new wife, Mary Lou, was a Jean Harlow-like peroxide blonde that he had picked out of the bongo world. They were living in Altadena just below the mountains, and the Riches and I were much excited when driving up there for supper. Our evening, however, was more muted than anticipated, with Mary Lou nervous about her cooking. Moreover, she was not a vamp but instead an early thirtyish woman extremely knowledgeable about art history. Dick often spent much of a lunch interval on a bench outside his Physics office reading *The Saturday Evening Post,* the weekly magazine that I had been brought up to despise. Its stories, however, were just what Dick wanted—devoid of any intellectual hang-ups and having predictably happy endings.

I was then reading *The Heart of the Matter* by Graham Greene, a story of a completely different sort set in Africa. Earlier I'd read his equally bleak, London-centered *The End of the Affair.* For his protagonists, love inevitably led to more agony than pleasure, and I easily identified myself with their struggles. I went through both with few interruptions, having been stricken by malaise that I worried might be mononucleosis. Although the Caltech students' clinic never confirmed this suspicion, I stayed largely confined to my Athenaeum room for much of a week.

When again I felt almost normal, Mariette Robertson came by the Athenaeum to invite me to a party at her house at which girls, as opposed to intellectuals and their wives, would be present. She lived beyond the Paulings in the foothills of Sierra Madre, at the end of a narrow driveway in her parents' expansive wooden house that was guarded by a nervous poodle whose curiosity made entering an awkward affair. As I was walking past Throop Hall several weeks before, Mariette had come up to me and introduced herself as Peter Pauling's former Caltech girlfriend. Like Peter, she was a child of Caltech. Her father Bob was a clever astrophysicist, who, like Linus, was one of Caltech's first graduate students in the early 1920s. Angela, her mother, was a Hungarian woman from the intellectual elite of Budapest that had produced their friend, Princeton's fabulous mathematician, Johnny von Neumann. Peter was still much on Mariette's mind, though she knew of his recent close friendship with young Nina, the Perutzes' Danish au pair, in Cambridge. Later, I could not deny Nina's beauty to Mariette, who now took comfort from Nina's return to Copenhagen.

Mariette's Saturday-night party, filled with very un-Caltech-like faces and bodies, went on into the early morning. Most intriguing was a tiny dark-haired girl who had painted a second pair of lips on her cheeks and whom I arranged to meet the following Monday after her classes ended at Pasadena City College. At the appointed time, however, she did not appear, and after an hour I knew I had been stood up. Soon, dates with Mariette on weekend nights made more sense than her staying at home with her mother and me imposing myself on married friends. But I had to assume that she would write to Peter, so I wondered how he would let me know.

9

Pasadena, Berkeley, Urbana, Gatlinburg, and the East Coast: March–April 1954

GEO GAMOW WAS by now back in Berkeley using two boxes of Fisher balls and plastic bases to build his version of the double helix. But he was stuck on how far the two chains were separated along the helical axis and he asked me to send him quickly the correct coordinates. With them he could finish his model before I drove north for a weekend with him and the Stents. Formal coding proposals, however, were his real love, and he went on to describe a new scheme using triangles that might be applicable to single-stranded RNA templates.

The week before, I had used my new car to move from the solitude of the Athenaeum to a three-room flat in a modest, single-storied dwelling on Del Mar Avenue. From there it was a quick morning's walk to Lake Street for a breakfast of orange juice, chocolate doughnuts, and coffee at Winchell's. Back to Caltech was an even shorter, five-minute stroll, punctuated by the calls of the mockingbirds that were the southern Californian equivalents of the robins that dominated the front lawns of my Chicago boyhood. By then, the Army was largely out of my mind because I officially knew that I could be part of the Public Health Service. But whether I would actually go to NIH would not be certain until the Presidential Review body made its decision. If they came down on my side, there might be good reason to stay at Caltech, especially given the real effort George Beadle had devoted to my cause. Also, spring had definitely arrived, and there were several semi-sunny days when the humidity approached East Coast levels.

The Biology scene, moreover, was enhanced by a stream of visitors escaping the last days of winter. The most exciting mind was the dashing Jacques Monod on a tour from France that had earlier taken him to Berkeley, where Gunther Stent hosted Jacques's playful intelligence. From Pasadena, the Delbrücks quickly whisked Monod to the desert and the opportunity to scramble up rocks that no one else could master. Jacques had first come to Caltech before the war, at a time when he was not sure whether he wanted to be a biologist or a musician. There was no uncertainty now, however, as to what Jacques sought from life. Until he discovered how bacteria could enzymatically adapt to abrupt changes in their food molecules, he would not rest. But whether he could get real answers in Paris before we learned how RNA was involved in the transfer of information from genes was unclear to me.

There was also a farewell party for Bill and Nora Hayes, who were returning to London after six months in the Delbrück environment. I felt awkward there because it was at my suggestion that Max had invited Bill to come, thinking I would do experiments on *E. coli* genetics with him once we had both arrived in September 1953. But once Francis Crick and I had found the double helix, what Bill was doing with bacteria bored me. All traces of this guilt were gone two days later when I picked up Leslie Orgel for a trip to Berkeley. As a theoretician, Leslie was not avoiding lab work and there was no question of bringing along his wife, Alice, as she was in the midst of a medical internship. Soon after we crossed the Tejon Pass, we cut to the west towards King City to avoid the trucks of U.S. 5. Leslie was usually oblivious of his surroundings but this day proved an exception. He luckily noted that my open Chevrolet convertible was on a collision course with a freight train whose tracks I had not spotted.

Gunther Stent's house on Channing Way in Berkeley, near the university, proved a more than adequate place to spend the night as well as the site of a Saturday-night dinner party with Geo. Earlier, we had gone to Geo's office, where he displayed an incomplete double helix that he had built using the wrong conformation for deoxyribose. So informed, Geo was not at all upset and later proudly sent me photographs of himself next to a much more normal-looking double helix.

Later, I diverted Geo from too many pre-dinner card tricks by proposing that he help Leslie and me form the RNA Tie Club. Its members would be united by their ties as well as a desire to understand the role of RNA in protein synthesis. During dinner, we all saw the value in limiting the club's membership to 20, one for every amino acid. Then it would neither be too big or too small. Already Geo saw a unique role for himself designing the club ties, tie pins, and stationery. By dessert, Geo had begun sketching the tie, which I argued should display RNA as a single-chained molecule. And we all agreed that while the RNA Tie Club needed to be clannish, it should not be a secret society. With luck, public monies could be raised to bring its members together. Obviously Geo, Gunther, Leslie, and I should be founding members as should

Berkeley, California, 1954: JDW with Inga Stent,
Leslie Orgel, and Gunther Stent

Geo Gamow with a model of DNA in Berkeley, spring 1954

Francis and Alex Rich. But who the others were to be was left for further deliberation.

I already knew that Geo and I would be spending some time together in the summer at the Marine Biological Laboratory at Woods Hole on Cape Cod. Geo was to be the houseguest of Albert Szent-Györgyi, the celebrated Hungarian biochemist, while I was to be there as an instructor in its well-regarded physiology course. Up to then, genes were not focused on in Woods Hole, where the emphasis had long been on embryology and physiology. Dan Mazia, the Berkeley cell physiologist then in charge of the course, was excited by the double helix and thought introductory lectures on DNA would set the tone for subsequent talks on how fertilized eggs develop into multicellular organisms.

Visiting Cold Spring Harbor this coming summer would not be a breather from girl-less Caltech. If only the Mayrs were to be there, I might count on Christa, although she was more likely to be around Boston now they had their New Hampshire farm. Even if Christa were elsewhere, however, perhaps the Woods Hole community was sufficiently large to give me hope that it would not prove another social desert. And, initially, I thought I would not be that far from Margot Schutt, my Henry James–fixated shipmate companion of the past summer. At last she answered my earlier letter of the fall. But she wrote mostly about friends we had formed on the Atlantic and said she was thinking of returning to Great Britain. I knew that I soon would be in Washington for a forthcoming National Academy of Sciences (NAS) meeting, so I proposed afterwards visiting Vassar College to see her. But she did not respond even though more than a month passed before I left Pasadena for my trip to Illinois and the East Coast.

During these days, Mariette Robertson and I kept each other from being lonely by going to movies and, once, going into Los Angeles for an "Evening with Beatrice Lillie." But this music-hall prattle proved embarrassingly boring, and must have left Mariette wondering why I remained so enchanted by things English. She was soon to leave Pasadena to accompany her parents to Paris, where her father, Bob Robertson, was to be the chief scientific adviser to the American military in Europe. Our friendship had even more reason to stay platonic, helped by her mother's poodle jumping on me whenever Mariette and I moved close while talking sprawled on her living-room floor.

After flying to Chicago, my parents drove me to Urbana and the University of Illinois to which Salvador Luria had moved soon after I finished my Ph.D. thesis. His departure from Bloomington was a big blow to science at Indiana University. Stupidly it never tried to match his good offer from Illinois, largely because of his leftist political actions, that included, unsuccessfully, trying to unionize its conservative faculty. At Urbana, Gunther Stent and I joined a group going to Gatlinburg, Tennessee, in the Smoky Mountains, where Alex Hollander had organized a mid-April meeting on genetic topics. Francis Crick would be there as would Max Delbrück. Eight of us from Urbana drove to the meeting in the large black hearse that Ed Lennox, an Atlanta-born physicist, now learning phage, used to move his many children to and from school. While driving south across Kentucky, Ed let Salva drive until we narrowly escaped a head-on collision as Salva played Russian roulette by passing on curves. Salva objected to his relegation to a rear seat, proclaiming that he had never been in an accident. But we had the premonition that Salva would have only one accident in his life.

No big surprises emerged from the several-day meeting, but it was fun seeing friends who had vanished from my life when I was in Europe. Moreover, the occasion let Max and Francis continue to disagree on whether the intertwined strands of the double helix would separate during DNA replication. At the meeting's end, Francis and I were passengers in a friend's car along the Skyline Drive into Virginia, where we caught a bus to Washington. Both of us were there to talk before a session on "Proteins and Nucleic Acids" arranged by Linus Pauling, who gave the opening talk. The smell of McCarthyism hung over the NAS and was only partially compensated for by the lush greenness of the adjoining parkland across Constitution Avenue. Francis talked about the double helix while I spoke about what Alex and I had learned about RNA and speculated on its role as an information-carrying molecule between DNA and its protein products. Later I went upstairs to meet the National Research Council staff member who interacted with the Selective Service Presidential Review Committee. To my delight, he told me that the Army might soon announce that they would no longer draft individuals 24 or older. As I had just turned 26, I was very unlikely to be forced into an Army that now focused on the still-malleable young.

At last my life seemed almost under my control as I pushed on to Swarthmore College to see Christa Mayr. During the next two days, we spent many hours wandering carefree through the wooded Swarthmore campus, then flooded with the first wave of migrating warblers. At other times, we stayed glued to radios tuned to the Senate hearings where Senator Joseph McCarthy at last met his match at the hands of the Boston attorney Joseph Welch. On the second evening we went into Philadelphia for dinner with a friend, who earlier had been with us at Cold Spring Harbor. The exciting unraveling of the McCarthy horror show dominated the evening talk, and I took pleasure in how effortlessly Christa held up her end of serious conversation.

At no time during these two idyllic days did I see a Christa about to fall into my arms, and I left the next morning for New Haven knowing that I wanted her more than she needed me. In New Haven, I was joined by Av Mitchison whose year with mice in Bar Harbor was going very well. His mother, Naomi (Nou), had already crossed the Atlantic to contrast Maine life with that of the Mississippi sharecroppers that she had visited in the middle 1930s. Like her brother Jack, Nou had long been an intellectual star of the left. Just before she flew back to Scotland for New Year's Eve, she and Av toured the Socialist world of New York City, finding its inhabitants too removed from political reality. Av's sister, Val, would also visit Maine, soon taking a month away from her job covering the Royal Family for the *Daily Mirror*.

Av and I sampled a lot of New Haven in three days. When I was growing up in Chicago I thought that the good manners of Yale were for sissies. But everywhere my Aunt Betty dragged me to we were received with genuine warmth and good food. In turn, Av delighted in telling strangers that he now worked in Bar Harbor with the very nice girl I knew at Indiana University. To my discomfort, he let on that she (Teke) might still be mine. But I knew my previous infatuation would never return and hoped Av would later tell her about my new love for a Swarthmore girl.

Pasadena: May 1954

ON MY RETURN to Caltech, I gave six lectures on bacterial genetics at the suggestion of George Beadle. The first was dreadful, but I used its failure to think through ways to lend the remainder coherence, if not actual zing. I was simultaneously preparing for my Woods Hole labs, where the students would do the now two-year-old Hershey–Chase experiment that showed that phage DNA, not protein, carries its genetic information. I persuaded Victor Bruce—Manny Delbrück's engineer-turned-biologist brother—to come and help me, arguing that he should know how Woods Hole stacks up against Cold Spring Harbor. Matt Meselson, then finishing his Ph.D. thesis under Linus Pauling's supervision, was coming, too. He was also unmarried and some weeks before had invited me to drive east to locate prospective girlfriends at Scripps Women's College in Claremont. But we came away empty-handed and talked science all the way back to Caltech.

As the spring came to a close, I saw George Beadle ("Beets") constantly on the move in Kerckhoff Biology Building to find out what he could do on behalf of his faculty. He sensed that I needed to be cut free from any further dependency on Max Delbrück to sink or swim on my own. As soon as the Army ceased actively to pursue me, Beets offered me the position of Senior Research Fellow in Biology to start on July 1. Going with this position was a salary increase from my fellowship's $3600 to the then-princely sum of $5000. Here Beets had the backing of Max, who nevertheless expressed disappointment with my progress over

73

these first nine months at Caltech. With almost no deliberation, I accepted Beets's offer, in part because he had worked so hard to keep me out of the clutches of the military. But I also knew that Caltech should be the perfect American place for me to pursue science if I had an appropriate Californian girlfriend. When I told Alex Rich of my decision, he was disappointed that I would not join him in Bethesda. He knew, however, that NIH and the Public Health Service would never have the same academic cachet as Caltech.

Beets's effort to keep good publicity flowing about Caltech led me to be approached by *Vogue* magazine, who wanted my photograph for their August issue. They planned to devote several pages to young celebrities who were making it big in the States, and I was to be included for the discovery of the double helix. The resultant publicity, I thought, should make "with it" American girls more eager to know me. No one at King's College London or doing biochemistry in Cambridge was likely ever to see this issue of American *Vogue*. In fact, the only scientist I knew who read the British *Vogue* was Francis Crick because Odile regularly purchased it to give style to their magazine table. A week later, a Hollywood photographer, who frequently shot for *Vogue,* turned up in the company of the attractive wife of an astronomer connected to the Mount Wilson Observatory. Talking to her brought animated smiles to my face, and they got the shots they needed.

At that time, I was working on the first draft of the second paper that Alex and I would write about RNA. Our first manuscript was soon to appear in *Nature* (May 22) and this new one was for *The Proceedings of the National Academy of Sciences* in an issue summarizing the talks presented the month before in Washington. Alex and I discussed its general form when we drove out to Joshua Tree National Monument for a camping trip with the Delbrücks. This was to be the Riches' last excursion to the desert before their return to the East Coast and Alex's assumption of his new job at NIH. To escape the weekend traffic we returned on Monday, stopping off in Palm Springs to explore its glitter and going into several fancy clothing shops inquiring about garden-party clothes. Jane tried on several frocks, trying to keep a straight face while I talked about future parties on the great lawns of the Cape Cod summer mansions near Woods Hole.

Initially I had planned to spend only the first half of the summer at Woods Hole, but by now I found it made sense to be away the whole summer. George Gamow would be there only in August. A late May letter indicated that he was still more than keen about the RNA Tie Club, writing that each of its 20 members should have a tie pin engraved with their respective amino acid abbreviations, such as GLU or VAL. Obviously, I must persuade Francis to join us at Woods Hole before he returned to England. There we could hold the first Tie Club meeting.

As I was to be away for three months, I decided to give up my apartment to Leslie and Alice Orgel. During my last weekend there, Mariette Robertson and I drove the freeways to Hollywood and wandered along the unromantic tackiness of Hollywood Boulevard before our movie began. Afterwards we went back to Del Mar Avenue to pack up my belongings for storage over the summer. Saying good-bye, our touching proved hard to stop. On the floor, we almost made our futures awkward but suddenly held back, knowing that each of us wanted someone else more.

As I drove back to Pasadena, I was concerned that, once in Europe, Mariette would not prove up to Peter Pauling's needs. His last letter to me complained that he had yet to meet any of the many rich, beautiful, and clever girls that Cambridge is famous for, though he did know some with only one of these three attributes and more with none. But if he was trying to stop getting along a little too well with girls, he had made the wrong car move. His brother, Linus Jr., had put at his disposal a recently acquired 1930 Mercedes-Benz open touring car, 18 feet long but mostly engine. The fact that it went only three miles for every gallon and could be used only for special occasions kept Peter from becoming the best-known person in Cambridge. His immediate concern, though, was that the May Ball in Peterhouse cost so much that not enough tickets had yet been sold. But at least he had a bright new girl to add to his car's splendor.

By now, the girls in Peter's life were a headache for John Kendrew, who had just written me again about his ambivalent feelings as Peter's Ph.D. supervisor. Personal charm had not helped Peter measure the absolute intensities of several key X-ray reflections from myoglobin crystals. This task should have consumed at most a month of time, but Peter barely completed it after nine months of capricious complaints. Giving Linus Pauling's son the ultimatum of work or get out, however, was not yet in the cards. Everyone enjoyed his light, evasive banter over morning coffee, and, given the tiny size of the unit, Peter's absence would be sorely felt. Moreover, John had a more serious worry that the unit would soon be kicked out of the Cavendish because of Sir Lawrence Bragg's impending move to London. Bragg had resigned his professorship to be the Director of the Royal Institution, a position earlier held by his equally famous father. The first thought of the newly appointed Cavendish Professor, Nevill Mott, was that the biologically inclined crystallographers should find a more suitable home. So far, however, no suitable alternative to the Cavendish site had been proposed.

Francis and Odile by then had a new daughter, whom, to my disappointment, they decided to call Jacqueline. To firm up his ideas while stuck in Brooklyn, Francis temporarily was keen to do a book for Academic Press, but, if completed, John Kendrew was sure that it would not be bought at King's College London. Its structural chemists were again hopping mad at Francis for what they considered a further poaching of their intellectual property. After reading a paper from King's on the structure of collagen, Francis concluded that Pauline Cowan had made a bad mistake and excitedly wrote back details of his newly conceived alternative model. Those at King's were not pleased. John Randall, its Wheatstone Professor, angrily wrote to Francis that "you will lose the respect of your scientific colleagues . . . ," and Maurice Wilkins blew steam to John for more than four hours. Whether Francis had hit upon the right answer, however, was not immediately clear. But on the good side, Maurice had told John "that in contrast to Francis, Jim had been carried away last year by the impetuosity of youth and later might turn out not that unbearable."

Cheered up by this unexpected assessment of my character, I made final plans for getting to Woods Hole. I would not be driving alone because both Leslie Orgel and Jane Rich opted at the last moment to join me, tempted by my offer to pay for the gas that would transport them to Chicago and New York City, respectively. Leslie wanted to look over the University of Chicago while Jane saw a way to be with her parents before joining Alex at the National Institutes of Health. All three of us were thus in high spirits as we headed east out of Pasadena along Colorado Avenue. Still then early in the morning, we soon would be on the main highway speeding across the Mojave Desert toward Las Vegas. Although the temperature would be stifling, the filthy smog would be gone.

11

Woods Hole: June 1954

THE HIGH DESERT temperatures kept us worried that our radiator water might boil over until we passed through Zion and Bryce Canyon National Parks. At the latter, we went off the paved highway for what we anticipated was to be a more scenic route. On the map was a tiny red line leading to the northeast that Leslie was particularly keen to traverse, wanting real wilderness unpolluted by tourists attracted to the multicolored limestone spires. There was no problem in reaching the neatly laid out little Mormon town of Escalante, but beyond it we were no longer on a paved surface.

As the road climbed through a ponderosa pine–dominated forest, it degenerated into a rutted track. Our average speed fell to no more than 10 miles per hour because of the precipitous voids that appeared to one side or the other. Luckily the constant turns gave me no opportunity to gaze downwards. It was virtually dark when, to our immense relief, a small sign marked the 9200-foot pass, some 3000 feet and two hours above Escalante. Optimistically, we thought the worst must be over, but only the absence of moonlight kept us from being petrified during our subsequent descent along a narrow ridge called Hell's Backbone. Two more hours of nervous driving passed before we reached the canyon floor where we spotted a ranch-like motel. Jane then did not want to pay for a bed when she could use a sleeping bag on the ground outside. Ungallant as it was, Leslie and I did not join her but forked out $5 each for the certainty that we would not have to drive the next day half asleep.

Jane in no way regretted her decision, sleeping so soundly that her distant twisted sleeping bag was thought by a stranger to be a sleeping calf. Later she was happy to pay for her breakfast, after which we traversed the less-demanding roads to Hanksville, and finally hit Green River and its transcontinental highway.

The remaining mountain passes proved child's play in comparison to those of our scenic bypass, and we were across the Rockies and into the plains of eastern Colorado before we stopped for the night. Crossing Kansas brought the excitement of water spouts, like tiny tornadoes sometimes, but afterwards there was little but corn to gaze upon until we reached Chicago. There Leslie left us to see the chemist Robert Mulliken, with whom he wanted to talk before deciding whether, after a year in Linus Pauling's orbit, he should spend equal time at the University of Chicago before returning to England. Ninety minutes later, after Jane and I had driven past the steel mills of South Chicago and Gary, we reached my parents' home beside the Indiana Dunes State Park, where I was again among the familiar songs of the phoebe and the red-eyed vireo. I had hoped we could then somehow link up with Av and Val Mitchison, then visiting Indiana University some four hours to the south. But they ran out of time, and Jane and I were off the next morning for New York. The next afternoon, I left her at her sister's apartment and was in Woods Hole by noon of the next day.

To most people, Woods Hole meant ferries to the islands of Martha's Vineyard and Nantucket, but for scientists it stood for the Marine Biological Laboratory (MBL) that was established there in 1888. Situated around a small pond that opened into Vineyard Sound, MBL's reputation came from its ready access to a plethora of marine animals useful for studying physiologic and developmental processes. Woods Hole was once a small village but its population now greatly expanded in the summer—through the influx of scientists to MBL and through the occupation of massive summer homes by many of America's wealthier families. Particularly impressive were the wooden shingled mansions on Penzance Point, a narrow, mile-long, highly curved sliver of land that originated just beyond the last of the MBL buildings.

Only one scientist had a home on Penzance. It belonged to Albert Szent-Györgyi, the Hungarian-born biochemist with whom George

Gamow would be staying during August. Called Seven Winds, it was one of the more modest houses, sited on the Buzzards Bay side, more than halfway to the tip. Through his Nobel Prize, awarded for the isolation and identification of ascorbic acid as vitamin C, Albert was the most famous scientist in Hungary and, after the war, had been seriously considered for its presidency. But the moment the communists seized power in 1947, Albert fled with his family. His political views were on the left, but his free-thinking mind was incompatible with the rigidity of communist orthodoxy. Initially he assumed that once in the States he would obtain an important academic position. But American university life and Albert moved to different beats, and his former patron, the Rockefeller Foundation, was unable to place him in conventional academia. So Albert established the Institute for Muscle Research at MBL, populated largely by fellow Hungarian refugees. By then, the annual budgets of the National Institutes of Health were rapidly rising, allowing Albert to get the grant support that covered research costs. It also provided him with a salary sufficient for his celebrity-like existence, made evident by the white Cadillac that he alternated with his dependable motorbike.

I first saw Albert's house late the afternoon of my first day in Woods Hole. After moving my belongings into the brick dormitory that overlooked Eel Pond, I had lunch in the still semi-empty wooden cafeteria. Then I went to the Old Mains, the turn-of-the-century laboratory where the Physiology Course had been taught for almost half a century. Already all the other instructors had arrived, and I found myself assigned a lower ground-floor lab. As Victor Bruce was not arriving until the next day, I had time to explore the grandeur of Penzance Point. To join me, I persuaded a young zoologist to abandon her lab chores for a tour of the big houses that were then still largely unoccupied. In the grounds of the largest—the white-shuttered opulence of one of the Pittsburgh Mellons—we most admired a large gazebo looking onto Buzzards Bay. Later, on the main road, we came upon Seven Winds, and its guest cottage next to a rocky beach, off which was moored a raft for swimming.

My first day ended at the Captain Kidd, which served food but where most people went instead for the beer served on several round tables that dominated the large room overlooking at its far end Eel Pond. Here

Albert Szent-Györgyi on his motorbike at Woods Hole, 1957

I was among biologists of many persuasions, who saw no reason to treat genetics as more than another important branch of biology. I quickly sensed an absence of Cold Spring Harbor–like intellectual intensity. But at the summer's start, I almost welcomed this slower pace, hoping to broaden my biological outlook to the basic facts of physiology, particularly those that related to electrical signaling along nerve cells.

Within several days, however, I was back into my past, making the two-hour drive up to Cambridge to see Christa Mayr, now home from Swarthmore for a summer job at Harvard. Victor Bruce came along, still somewhat worn out by the journey from California with his wife, Nancy, and their year-old baby boy. On the second floor of Harvard's massive red-brick Biological Laboratories, dating from the early 1930s, I found

Christa helping with *Drosophila* experiments that Paul Levine, a newly appointed Assistant Professor, hoped would lead him to a tenured position. Like most summer jobs, Christa's was far from intense and we saw no hurry to move on. But after she expressed enthusiasm about later spending a weekend at Woods Hole, the time came to leave. Saying that we were expected in Paul Doty's nearby chemistry lab, we went down the stairs and in less than two minutes had moved from the boredom of old-fashioned biology into the sparkle of chemistry well done.

Even though Doty's labs were located within the turn-of-the-century Gibb's building, they had been modernized—in contrast to those of the Bio Labs, which reeked of thirties mustiness. He was keen to show that double helices fell apart into intact, single strands when exposed to conditions that severed the hydrogen bonds holding together the adenine-thymine and guanine–cytosine base pairs. This much more confident Paul, then aged 34, already had tenure, having come to Harvard six years before after a brief spell teaching polymer chemistry at Notre Dame. Being here was a big plus for Paul because Harvard's Chemistry Department had no equal in the States, if not the world. Year after year, his incoming graduate students were as good as they come. One, in particular, Helga Boedtker, proved particularly important; Paul married her after his first marriage had fallen apart. At first, they lived on the top floor of a nearby 1920s apartment building, and Paul took us there for lunch sandwiches. There I got real gossip about Harvard and its Chemistry Department filled with too many stars ever to be dominated by a Linus Pauling, popelike figure. In contrast, I gathered that Harvard's Biology Department was floundering in the past, with its few outstanding professors such as Ernst Mayr and George Wald totally outnumbered by one mistaken appointment after another. But Paul held out hope that McGeorge Bundy, the young new Dean of the Faculty of Arts and Sciences, was too bright to let biology continue its dreary path to nowhere.

Harvard thus loomed large in my mind as we drove back to the Cape. A faculty position there could give me not only a chemical colleague focused on DNA but also the opportunity to be near Christa and, if not her, the many girls whose faces caught my eye as I walked about Harvard Yard. With that end in mind, I returned five days later to the Boston region to see Lee Wakefield again and learn of the post-graduation plans of her and her elusive Vassar roommate, Margot Schutt. Alas, neither would be anywhere near Boston over the summer. Nor would Lee be visiting her mother's family on Naushon Island off Woods Hole because she had opted for a summer of outdoor life in Wyoming. My summer fun would have to come from the world of science.

12

Woods Hole: July 1954

THE PHYSIOLOGY COURSE at Woods Hole initially took up all my time, with lectures in the morning followed by labs that ran through the afternoons. Harvard's George Wald gave the most polished talks, addressing the chemistry of vision. His Brooklyn origins were utterly absent, except in the jokes he later told in the security of his home. In contrast, there were no big crescendos during the lectures on nerve cells by the more modest Steve Kuffler. They were given with such good will that I could never tell him that I remained ignorant of how electrical signals move along nerve fibers. Later the theatrics of Albert Szent-Györgyi, whose guest appearance mixed up apparent wisdom on how muscle proteins work with thoughts on magic molecules that block cancer cells from dividing, provided genuine fun. Those in Albert's lab, however, knew his enthusiasms often exceeded his facts, and that for the moment there were no rabbits in his hat. At the other extreme were the ponderous orations by Max Lauffer on oxygen-carrying proteins. His labs were equally dreary, leading to an afternoon crisis after several of us temporarily absconded with the experimental lobsters.

We were encouraged in such antics by Albert's much-younger distant cousin Andrew Szent-Györgyi and his super-attractive wife Eve. Like Albert, both had fled Hungary when the Russians took over and were now working together under Albert's patronage. As year-round inhabitants of Woods Hole, they let me know the local gossip and what to expect of dinner with several boring married couples. As a rule, the most

attractive girls invariably took the invertebrate course, which emphasized dissection and drawing, rather than the physiology course.

On the plus side, the students who opted for my phage section knew
how to do experiments and got the right answer—that phage DNA, not
phage protein, carries the genetic specificity. Earlier I had felt rejected
when Frank Stahl, then working on phage for his Rochester Ph.D.,
chose not to take my lab. But he did not want to waste time repeating
techniques he had already mastered. The lab he opted for then failed to
inspire him and, at the end of most afternoons, he sat in front of our Old
Mains lab drinking gin martinis. Frequently he was with Matt Meselson, who had just arrived from Caltech and was getting his first exposure
to non-gene-dominated biology.

At my suggestion, Matt had brought with him some RNA to do titra-
tion experiments aimed at showing whether the bases of RNA were
tightly hydrogen-bonded. Not many afternoons, however, were needed
to show they were not, and Matt's summer largely became one for talking about what he should do after he completed his Ph.D. under Linus
Pauling. High on his possible objectives was finding a way to distinguish
newly made DNA strands from their parental templates. In theory, they
could be distinguished by the rate at which they sedimented in an
ultracentrifuge using stable carbon and nitrogen isotopes to differenti-
ate parental from progeny chains. But the differences would be so small
that only very cleverly used ultracentrifuges might give clear answers.
So I urged Matt to consider post-docing in Sweden. The ultracentrifuges had been pioneered there, and its women were said not to have
sexual hang-ups.

After the course started, instructors and their wives were invited
after dinner to the director's home for not-even-slightly-alcoholic Friday
night punch and dessert. The occasion was even heavier than I feared,
and I quickly left to search out the Bruces to see if Nancy had finagled
a last-minute invitation for me to a Saturday lunch they were going to
in a big house beyond Nobska Beach. Their invitation had come from
the mother of one of Nancy's Vassar friends, and the lunch was likely to
be a class act. It produced, however, more moments of good manners
than of unexpected conversation or beauty. Our hostess's children had
not yet arrived for the summer, and the gathering ended much sooner

than I wanted—I knew our Old Mains lab would be deserted and I would have six hours to kill before I could move on to an evening party at J. P. Trinkaus's home on Devils Lane. Trink, as he was known to all, was an embryologist at the Lab and seemed to know everyone. No one could complain that his beer-drenched parties were staid or that the couples that came together always left together. But after staying too long that night, I realized that no more came out of Trink's parties than went into them.

On weekdays, however, most of us tried to maintain the facade that we were in Woods Hole primarily for science. But I deeply offended several old-timers by giving lectures in unlaced tennis shoes and wearing my floppy hat at night as well as during the day. My water pistol was also judged inappropriate, even though I generally restricted its aim to a pretty girl from the South taking invertebrate lab work too seriously. But I saw no reason to dress or act differently than I had during past Cold Spring Harbor summers when pomposity always backfired.

As July progressed, I gravitated more and more towards those few visitors who had an interest in genetics. Particularly rewarding to have about were Boris Ephrussi (co-author of the *Nature* hoax of the previous year) and Harriet Taylor, who had just arrived from Paris. Later they were moving on to Harvard, where Boris was a visiting professor for the fall term. We were often joined at meals by the New York–based geneticist Ruth Sager and her husband Seymour Melman, an economist at Columbia University seemingly resigned to the fact that his wife bristled when identified as Mrs. Melman. Ruth, then at the Rockefeller Institute, had worked at Columbia and was full of gossip about the controversial German scientist, Franz Moewus, long suspected of faking much of his work at Heidelberg. There, he and his wife had used the green alga *Chlamydomonas* purportedly to show how genes control the synthesis of enzymes. In an attempt to establish Moewus's innocence, Francis Ryan had brought him temporarily to Columbia to repeat certain key experiments. Now Moewus was in Woods Hole with his wife to be part of the Botany course. So far the jury was still out.

With the exception of a brief note to my parents, I had not written a letter since arriving in Woods Hole and by mid-July had little reason to find envelopes addressed to me in the mailroom. To my surprise, I found, forwarded from Pasadena, a formal invitation from the Honor-able James and Mrs. Griffiths to the wedding, several days later in Lon-don, of their daughter Sheila to Roy Pryce, the young historian she had met during her last days in Rome. Although I had long stopped thinking about Sheila since our meeting in the Italian Alps during summer 1952, the invitation jolted me.

In the afternoons, Matt Meselson and I frequently retreated to a large garden that ran down to Quisset Harbor and whose largely absen-tee owner permitted the occasional visitor to enjoy. Two summer village waitresses often drove there with us, one with long blond hair while the other was slightly plump. They were in search of more structure for their lives than Matt and I wished to give, particularly when the afternoons were over and there were evening talks to attend, if nothing better came along. There was a drive-in movie theater two miles away in Falmouth, but I was never able to persuade the pretty little blond girl in the adja-cent invertebrate course lab to come along. So I consoled myself with the knowledge that the social scene was bound to improve. Not only would George Gamow soon arrive, but Francis Crick's job in Brooklyn had effectively ended, giving him the month of August before his return to England. And, Christa Mayr was coming down to Woods Hole for the last weekend in July.

Before Christa got off the bus from Boston, I was apprehensive that she might have found a summer boyfriend while working at Harvard. But the enthusiastic way she had recently confirmed the visit gave me reassurance that no one new had seriously entered her life. For Saturday night, I found her a vacant dormitory room and we momentarily went to it after her bus arrived. From there, we walked on to the younger Szent-Györgyi's house for I wanted Albert and Eve to meet her. Quickly Christa was let in on the practical joke that we hoped would mark the high point of this Woods Hole summer. With both Francis and Geo about to arrive, the time had come for a super summer party. And to give it some style, we hit on the idea of sending fake party invitations to the entire Woods Hole community. The invitations would come, reputedly from Geo himself, to celebrate his arrival through a "Wiskie-Twistie" RNA party at the Szent-Györgyi guest cottage. Albert's wife Marta would not welcome the impending mob scene, but if Geo later went along with our gag, she could hardly stop him being himself. Marta's style was that of a Hungarian Countess, and already I had fallen into her bad graces by being too casual at a barbecue presided over by Albert for his younger Hungarian colleagues. My standing with her could hardly have been any lower, so I felt no harm could come from being identified later as the hoax's originator.

It became even more apparent to me that a real party was badly needed when Christa and I ended the day at the Trinkaus's weekly beer and gin bash. There, to my annoyance, Christa enjoyed the attentions paid to a pretty new face, while I was part of conversations warmed over too often. The next day, however, I had her undivided attention as we canoed across the Hole to Nonamesset, the closest of a string of unspoiled Elizabeth Islands lying off Woods Hole. Soon we were walking through a sheep meadow towards the narrow bridge going over to Naushon, the largest of the islands, on which most of the Forbes summer homes had been built. Nonamesset now belonged largely to sheep and their attendant ticks that we were constantly pulling off our bare legs below our swimming suits. Upon crossing the bridge, we continued on to the secluded harbor below the multistoried wooden mansion of Cameron Forbes, the 90-year-old family patriarch who long ago had been governor general of the Philippines. No cars were permitted on Naushon and its outlying appendages and horses and buggies used to go between the houses. Although we were trespassing, we didn't believe anyone would mind and had no hesitancy in waving to the several carriages that passed us, presumably taking Forbeses to Sunday lunches at each other's homes.

After Christa went back to Cambridge, I joined the Harvard behavioral biologist, Don Griffin, for a tern-collecting trip to a deserted island off the Maine coast. Don, a newly appointed professor in the biology department, had asked me to come along after learning that I had once planned to be an ornithologist. After the terns were collected and banded, they were to be taken some 100 miles inland to see whether, upon release, they would quickly fly back to their home island. The chance to be on a Maine tern island by itself would have led me to the long car ride. More important, I thought, Don might start thinking of me as a potential member of the Harvard faculty.

The weekend was about to start when I got back to Woods Hole after three days away, and I found a letter from Sydney Brenner telling me that he would soon visit the lab. He was away from Oxford for two months doing experiments in Milislav Demerec's lab at Cold Spring Harbor as well as taking the bacterial genetics course there. His wife May had remained in England with their son Jonathan, largely because of her distaste for the American political scene but also because Sydney's stipend as a graduate student was not meant to include bringing a family along.

Francis Crick had already arrived and, with dormitory accommodations chancy to find, rented a small room in the home of long-time summer residents, the Littles, who had seven children known to all as the "little Littles." Geo was still two days away, coming with his wife, Rho, and their almost out-of-adolescence son, Igor. Early on Saturday morning, with the help of Andrew and Eve, I secretly mimeographed more than 100 copies of the invitation to Geo's party for later placement in the mailboxes used by Woods Hole scientists. Near midnight on Sunday, I placed the invitations into the various pigeonholes. For 48 hours the MBL community would savor their unexpected invitations. Then the real fun would begin.

13

Woods Hole: August 1954

THE EUPHORIA INITIALLY generated by multitudinous Seven Winds invitations slowly turned into a search for the culprits who had set up the denizens of Woods Hole for their subsequently deflated egos. Francis Crick was among those taken in, cheerfully asking me whether I had also been invited and anticipating an evening of Gamow-promoted high jinks. George Wald was mad at being had and stopped me in front of the Captain Kidd with the rumor, soon turned into fact, that Gamow denied any input into the invitation bearing his name. Although Wald did not directly imply that I was one of the pranksters, his disapproval indicated that I might have to think about a girl-containing university other than Harvard. Even deeper disapproval came from the elderly Guternatches, whose pre-Nazi German origins gave them the unshakable belief that students did not trifle with their professorial elders.

Rho Gamow, meanwhile, told everyone she knew that Geo was a co-conspirator. In more than 20 years together, Geo had never been happier than when perpetuating a practical joke. I knew, though, that I was in imminent danger of being thrown, clothes on, into Eel Pond by the Captain Kidd habitués who persisted in believing that DNA was not the only way of looking at biology. Such possible humiliation happily passed when Geo announced that the big party would go ahead and that all who had received the phoney invitations were welcome.

Those of us who knew Geo were already part of a semi-continuing party atmosphere rotating between Seven Winds cottage and the Captain Kidd round tables, where Geo knew that he could find an ever-changing set of faces unprepared for his limericks and card tricks. Particularly lethal were the drinks that Geo prepared, for his idea of a tall drink was a tall glass completely filled with whisky. Whenever possible, I made my own drinks and so avoided having to slip outside to empty my glass on the rocks going down to the water. By now, Sydney Brenner was briefly with us, having decided to interrupt his Cold Spring Harbor stay to be with Francis and Geo.

With the real party still a week away, Francis, Sydney, and I joined Geo daily in the water-facing living room of the cottage for extended discussions of Geo's genetic codes. My mathematically deficient brain meant that I often missed the validity of the argument either for or against Geo's diamond code or the alternative schemes hatched by Geo and his close friend, the bomb-making physicist, Edward Teller. We badly needed much more data on which amino acids were next to which in the newly sequenced proteins. We had the insulin sequence recently completed in Cambridge by Fred Sanger and that for the peptide ACTH (adrenocorticotrophic hormone). Even with the limited data on hand, the diamond code looked dicey, irrespective of the implausible stereochemical assumptions it placed on how amino acids would be recognized by triplets of base pairs.

After a third afternoon of intellectually forward and backward moments, Albert Szent-Györgyi asked us up for drinks at Seven Winds itself. There I had to face a Marta silently mad about the hordes that were to descend on her principality in a week's time. Equally standoffish was her college-aged daughter, Ursula, who, to my annoyance, only smiled when Francis caught her eye. My attempts to mollify Marta obviously went nowhere because the next morning she had placed in my mail slot a small envelope filled with sand. Its accompanying note asked me to return the contents to its proper location. My golden tennis sneakers, that had happily reappeared painted after several days of stolen existence, must have annoyed her. She now thought them the source of the sand defacing her living-room rugs.

After the fifth day, the coding powwows began to falter as it became painfully clear that further discussion was impossible without more amino acid sequence data. Nonetheless Geo's morale remained high.

He was especially proud of a driftwood model of the double helix that he had put together outside the cottage. It used my golden sneakers for its base and my floppy hat for the top. Sydney Brenner had already left by then, for he was needed back at Cold Spring Harbor to give a Friday-night lecture on the mutant bacteria he had isolated during his summer's stay. But Francis's visits to the Seven Winds, if anything, became more frequent, using a girl's bike belonging to one of the "little Littles" to have afternoon swims off the cottage with Ursula Szent-Györgyi.

On Sunday, Francis and I were invited to lunch off Strawberry Lane at the summer home of Dorothy Wrinch, a formidable white-haired English mathematician then on the faculty of Smith College in Northampton, Mass. In the late 1930s, Dorothy was notorious for her "cyclol" model of proteins. It proposed an interlocking caged structure as opposed to the already then favored model of a folded linear polypeptide chain held together by hydrogen bonds and Van der Waal's interactions. Linus Pauling thought the cyclol model absurd, violating everything the rules of chemistry stood for. Nonetheless, for several years, the cyclol model remained on the scene because Dorothy had a backer in General Electric's famous physicist, Irving Langmuir. But by now cyclols were long dead, Pauling's α-helix model was triumphant, and Dorothy had become less prickly and more fun.

Most of Dorothy's luncheon guests were familiar MBL faces with one sparkling exception—Ellen, a fetching, green-eyed, red-haired girl in her early twenties, whom I had seen earlier in the box office of the Falmouth Summer Playhouse. Soon we began exchanging facts about ourselves and how Francis, Geo Gamow, and I were working out how genes provide the information needed to assemble amino acids into proteins. All too soon, Francis identified me as the sneaky perpetrator behind the fake invitation to what now could be the party of the summer, particularly if our new red-haired acquaintance would grace its presence. When she showed no hesitation in accepting Francis's invitation, I almost stopped worrying whether August 12 would be a success. Although there was a modest wedding band on Ellen's left hand, she gave no indication that she would be coming other than alone.

Until the last moment, we did not know whether Albert and Marta would boycott the party by not coming down from Seven Winds to the cottage at the water's edge. Meanwhile, Geo and I decided to share the cost; Geo provided the strong drinks, and I the beer, all of which we purchased at the liquor store near the ferry terminal. Virtually everyone invited came, except the Guternatches. George and Francis Wald arrived on time as did Ruth Sager, who earlier had virtually boxed my ears over her tongue-in-cheek invitation as part of Mr. and Mrs. Seymour Melman. She and Seymour walked in with the Ephrussis just in time to watch Albert and Marta enter and talk to Rho. Quickly the abundant whisky began dissipating into the countless tall drinks that Geo used his prerogative as host to dispense to all newcomers. Soon the noise level from the 100 and more guests inside the cottage precluded real conversations, and many of us moved to the grass between the cottage and the rocks at the water's edge. I went outside soon after Marta walked in, fearing that if I chatted too long with her, I might be forced to apologize for the sand on her rugs. I was also keen to spot our box-office beauty, who had promised to come the moment the play had started.

Happily, not too much more time passed before Ellen came through a gap among the long line of cars tightly packed outside the Szent-Györgyi compound. Even in the half dark she was a looker and in ordinary company should never want for attention. But if not immediately cared for, she might feel unwelcome in the MBL world—so long numbed by the absence of beauty that I feared it might not spontaneously ogle. I quickly placed a bourbon-and-Coke-filled glass in her hand and led her triumphantly towards Andrew and Eve Szent-Györgyi to show them that the right girl could take my thoughts away from Christa. A glance towards Francis reassured me that he would not quickly come over and dominate the conversation. His animated laughter, first directed towards fellow scientists, was pitched toward Ursula Szent-Györgyi, who had come in with her mother.

Initially, Urs made dutiful, insipid girlish conversation with Igor Gamow, then keen to be a ballet star. Before the Gamows' arrival, Albert and Marta hoped that Igor's presence might cause Urs to come to life and stop acting as if life was meaningless when placed in the company of her stepfather's colleagues. But, after five minutes of Igor's ballet ramblings, Urs could take no more. Happily, she now coyly bantered with Francis, long educated to the fact that good parties are for the amusement, not the education, of the fairer sex.

Urs had figured that her mother would bolt the occasion after not more than 30 minutes. But to her surprise and later annoyance, Marta passed more than an hour chatting happily before a momentary lull in the conversational roar made her suspect that the best moment of the evening had passed. It was best for her to announce her departure before there was no one on hand to notice. Quickly she gathered up Albert—always happy talking about fishing on the "middle grounds" or his daily swims around Penzance Point—and then Urs, who knew this was not the occasion to announce her independence from her mother.

At least one too many of Geo's tall whiskies consumed while talking to Urs was now reverberating around Francis's head. His conversational force was already ebbing when he made his way over to where Ellen and I had been earlier joined by Igor, whose ballet dizziness in no way implied disinterest in pretty girls. He needed no alcohol to come alive and his long Russian hair virtually fell into her face as he spoke of being a ballet student in a class run by a son of the great Russian ballet dancer and choreographer Michel Fokine. Jokingly, Igor recounted fears that heredity had given him his father's ability to dance while his mother, once trained for the ballet in St. Petersburg, had passed on to him her minor talents as a physicist. Like his father, Igor liked to have a good time. By now, however, the cottage was no longer the place to have one. The beer and whisky were gone, and those who wanted more to drink had to get back quickly to the Captain Kidd before it closed.

Sensing that I might unnecessarily lose my beauty to the dull security of her own car, I quickly got an increasingly hung-over Francis into a departing vehicle that would let him off at the "little Littleses'." Igor, Ellen, and I were then free to do a quick mop-up of the plastic glasses and beer bottles that had been surreptitiously scattered on all sides of the Szent-Györgyi cottage. The task completed, Igor gallantly volunteered to help Andrew and Eve take charge of the chaos still within the cottage, allowing Ellen and me to slip away for a half-alcoholic, hand-holding walk beyond the big house at the end of Penzance Point. On a bench at the tip we watched the tidal water race through the hole, and she tried to be matter-of-fact in talking about a husband back in Boston. Things weren't working out well, she said, and she was in Falmouth for the summer to sort out her thoughts and give a more artistic lilt to her life than had yet come from her briefly married existence.

To keep the evening alive, we went on to explore the large unoccupied shingled house next to Seven Winds, which could be entered through an unlocked kitchen window at the rear. Its owner, for tax reasons, did not seriously try to rent it and used it instead to store surplus furniture unneeded in his even-larger summer house on the other side of Falmouth. Near twilight I had earlier broken into it several times. Again this evening the kitchen window remained unlocked. Not ready to end the fog-filled after-midnight, Ellen showed no hesitation in also climbing over the window ledge, with my warning her not to disturb the kitchen utensils. Displacing them might later alert the owners to the fact that their house was not secure. Downstairs we were unnecessarily nervous in front of the large windows that looked over the water, but upstairs as we moved hand-in-hand from one bedroom to another we were beyond observation. Only when again in the comfort of a large downstairs sofa did we move from hand-holding to embracing.

Unfortunately, my alcoholic haze lifted enough to realize that my incredible good fortune of finding Ellen not only pretty but willing was occurring less than 10 hours before I was scheduled to drive to New Hampshire for a weekend at the Mayrs' newly purchased farm. Now I would have given everything for my farm visit to be postponed. My momentary physical hesitation gave Ellen the time to express her fears that we were going beyond her point of no return. And, in turn, I awkwardly revealed my long-hoped yearnings for marriage to the daughter of a Harvard professor. The time to climb back over the kitchen sink had come and, after rearranging the sofa cushions, we went out into the just-breaking dawn and back to her car. Then she was gone, perhaps irretrievably. Only the alcohol in my blood kept me from immediately regretting I had rejected an immediate gem for one that might never be mine.

Woods Hole, New Hampshire, and Cambridge (Mass.): August 1954

WHEN I AWOKE, Friday morning was half over, and the half-empty physiology labs revealed that I was far from alone in missing breakfast. Those there were working only halfheartedly, and I sensed that the past evening had been an unqualified success by mixing groups that before never saw reason to come together. Over coffee Eve reassured me that it was the best party she had been to since her student life of Hungary, and she saw no reason why I had to be blackballed by George Wald from a subsequent appointment to Harvard.

The thought that I could use my talents as an impresario, if my career as a scientist began to falter, filled my head as I drove away from Woods Hole towards New Hampshire. After stopping for a late lunch at Harvard with Paul and Helga Doty, I was back on the road to the Mayrs' farm. There I found Ernst and Gretel tickled pink with their good fortune in having a country place for weekend escapes and the summer solitude that would facilitate Ernst's future writings on evolution. First, however, the farm had to be brought to the clean simplicity expected from their Bavarian origins. Painting the farmhouse red fell into Ernst's domain. Soon after my arrival, I offered my assistance. Smiling firmly, he refused, letting me know that I would have a full-time painting job the next day. I was thus free for Christa and her sister Susie to lead me around their largely forested acres that included a small pond for which I had been forewarned to bring a swimming suit.

Over dinner we bantered whether Ernst had enough clout to get me a Harvard offer. I slept well that night, sensing that Christa seemed as keen as her parents to have me part of their social scene. The next day I was a house painter until early afternoon, when, using the excuse of bird-searching, Christa and I hiked in the direction of a small mountain that loomed over the horizon. Only the next day did we go swimming in the pond, which was small enough for us to touch naturally when treading water in front of Ernst and Gretel, who soon came down to gossip.

Then we were all off to Cambridge from which I reached Woods Hole two hours later. Geo Gamow was to be there for several more days but already his moment had passed. All attention was on Franz Moewus, the German geneticist, then in Woods Hole. At last he had been definitively caught faking experimental results. Serious allegations against Moewus had first emerged 15 years before, when Av Mitchison's uncle, J. B. S. Haldane, published a brief note saying that Moewus's experimental results did not show the random variations expected from Mendelian-type genetic crosses. In reply, Moewus countered that he must have subconsciously selected for publication the crosses that best approximate to the statistically expected Mendelian values. There was also the question that the volume of experiences reported seemed far in excess of what one scientist and his wife, given their relatively modest academic appointment, could have carried out. On the other hand, if correct, Moewus's work on the genetics of the molecules underlying sexuality in the green alga *Chlamydomonas* had to be judged among the most significant genetic feats ever carried out.

The most straightforward way to rule out potential fraud is for an independent investigator to repeat the claim. This was why Moewus had been asked to help teach the Marine Biological Lab's botany course. In that capacity, others could be given his strains of algae and several crucial experiments repeated. But when hitches invariably developed Moewus put the blame on his fellow scientists not knowing how to culture his algal cells correctly. Then Moewus personally did a crucial experiment in front of several MBL observers, who later concluded that he likely used cyanide to make key cells immobile. In spite of all this,

Moewus had just given a prestigious Friday Night Lecture repeating claims that could not be reproduced.

Most keen to find Moewus honest was Tracy Sonneborn, who had championed his work since the mid-1940s. The week before Tracy had come from Indiana and was in low spirits over lunch with me, Ruth Sager, and Boris and Harriet Ephrussi. In our eyes Moewus had irretrievably blown his last opportunity to prove his innocence. Tracy could only wonder how he had been hoodwinked for so long, even years after Moewus's immediate German colleagues had lost faith in him and seen that his academic position at Heidelberg ceased. They still felt so strongly that a German scientist got up after Moewus's Friday lecture and bluntly stated that the results Moewus reported were not believed by those German scientists who knew him best.

I also had reason to be depressed by the outcome because five years before I had written a long-term paper for Tracy on the importance of Moewus's claims. Instead, I was now basking in the notoriety accompanying my appearance in the August issue of *Vogue*. On the same page with Richard Burton, I was described as having "the bemused look of an English poet." Francis Crick pointed out my previous publicity-averse posture but did not seem too upset. After all, none of his friends in England would ever see the issue. It was quickly spotted in Bar Harbor, however, by my girlfriend of my last Indiana University year. She used the occasion to write me that she was marrying a banker and moving to a small town in New Hampshire. The warm tone of her letter made me feel good because I felt badly about the awkward way I had backed out of her life.

As the summer visitors to MBL increasingly returned to their academic homes, I began to look forward to the phage meeting that was to be held during the last days of August at Cold Spring Harbor. Francis accompanied me there because the occasion would expose him to ways in which phages might be used to explore how DNA functions as the gene. The meeting's high point was Sydney Brenner—a substitute for Seymour Benzer, who was in Amsterdam—explaining the elegant genetic trick Seymour had used to map phage mutations at very high resolution. Afterwards, Francis went north to the annual Gordon Con-

"A scientist with the bemused look of a British poet":
JDW aged 26 in Vogue *magazine, August 1954*

ference on Nucleic Acids and Proteins. I wasn't invited but did not
mind because I was keen to return to the Mayrs' farm now that Christa's
summer job had ended, and she would be there for the week before
Labor Day.

Sydney came with me to New Hampshire. On our way, we stopped in
New Haven where my Aunt Betty saw that we were well-breakfasted
before going on to Cambridge. There we used a key, given by the
Ephrussis, to get into a flat on Trowbridge Street that was being reno-
vated for Boris's arrival at Harvard. We found the flat in shambles, but
we managed to sleep well before workmen burst in early the following
morning. A torrential rain was coming down and howling winds told us
that we were being hit by the big hurricane Carol that my aunt had forewarned us about. Never having been in one before, we carefully ven-
tured into the warm, windy rain searching out a coffee shop for
breakfast. Afterwards, I snaked the car around fallen branches and the
occasional uprooted tree to find that Paul Doty's lab had lost its electric-

ity. With no experiments possible, we talked science waiting for the storm to move on. By early afternoon the winds were no longer ferocious, and we drove over virtually traffic-free roads to the Mayrs' farm.

The next day, to let Christa meet Francis again, we drove some 60 miles north to the New Hampton School site of the Gordon Conference. On the way, in the tiny town of New Boston, we stumbled upon the Gravity Research Foundation—the bizarre brainchild of a wealthy investment adviser, Roger Babson, whose fame came from predicting the stock-market crash of 1929. In the foundation's nineteenth-century offices, we learned about Babson's obsession with the force of gravity that he held responsible for the childhood death of his eldest sister while swimming. Now he wanted to use his fortune to find ways to insulate humans from the harmful effects of gravity and towards that end sponsored an annual $1000 prize for the best 1500-word essay on ways to find new alloys that would reduce the strength of gravity. That this hope went against every serious physicist's idea of the nature of gravity did not faze Babson. He had even chosen New Boston as the site of his institute to avoid it being destroyed if an atomic bomb hit the Boston-area site of his Babson Business Institute.

We had to chuckle when reading its various pamphlets that had kooky titles such as *Varicose Veins and Gravity* and *Trucking Costs and Gravity*. Our arrival in New Hampton found us still in high spirits as we spotted Francis talking to a group of largely chemists displeased that the double helix had been discovered without their participation. In fact, behind his back several scoffed openly about using physical force to keep Francis in check. In contrast, Christa was charmed by Francis's conversational onslaught and so reported to her family when we arrived back for supper.

After dinner back at the farm, I found Christa not anxious to sleep and after a long walk down and back the country road beside their house, we started kissing in the darkened hall outside her room. When she finally went through the door, I was intensely relieved and fell asleep quickly. The next day we were quietly a couple and she accepted my invitation to come down to Woods Hole the next weekend for a chamber-music festival in nearby Coonenessett.

The Cape Cod she saw on her arrival was much changed from that she had seen in mid-summer. All the grass had turned brown, killed by the salt spray of Hurricane Carol. Large overturned yachts littered the inner shore of Penzance Point, and all the MBL ultracentrifuges had been put out of action by the several feet of salt water that had poured into its basements from Eel Pond. Moreover, MBL was now in double jeopardy from a new hurricane coming up the coast that might hit Woods Hole the next day. The thought of it dominated the early pre-concert supper at Andrew and Eve's cottage where Christa spent two nights of her visit. The cottage had just been brought back into order after the onslaught of Carol, and they did not anticipate another round of soaked belongings. That evening I also found Christa in slight retreat from the open affection of our last night at her parents' farm. The next day brought strong winds but not the mighty flooding waters of Carol, and Saturday night's opera by Gluck came off untouched. By Sunday afternoon Christa was again unrestrainedly sociable and highly enthusiastic about the Dvořák piano quintet that culminated the music festival.

Back in Cambridge the following day, Sydney Brenner joined us from New Haven where he had spent the week with friends he had made on his ocean crossing. All the Mayrs were on hand to see us start our journey across to the West Coast, which Sydney would visit for a month before going back to England and then to his family, who were by then in South Africa. At our parting it wasn't possible for me to kiss Christa good-bye, and there was a hollow feeling in the pit of my stomach as I had my last glance of her through the car window.

15

Northern Indiana and Pasadena: September 1954

As SYDNEY BRENNER and I drove as fast as possible across the flat dreariness of northern Ohio towards Indiana in late August 1954, I increasingly thought of my parents. For the last year Mother had been preoccupied with her own mother's failing health. At 93, it was not surprising that senility had enveloped Nana. But Mother increasingly found it difficult to cope with her incoherent rage and for almost a year she had been in a nursing home an hour away. So it came as a relief when I learnt on our arrival that Nana had just died and was to be buried the next morning. Not only had there been the emotional strain of the frequent nursing-home visits, but Nana's yearlong bills already had exhausted my parents' limited financial resources. Fortunately I was spending less than my salary and was able to cover the funeral costs.

The following day at the cemetery, and afterwards at their home, I shared memories of Nana with her Gleason-descended relatives. They lived in Michigan City near the land that Michael Gleason farmed after his arrival from Ireland at the time of the Great Famine. They were Mother's only relatives as her father had left Scotland alone and later was to die from a runaway horse accident on Christmas Eve when she was only seven.

At home that evening, I finally admitted to my parents my love for Christa. They quickly sensed that waiting until she finished college three years hence would not be easy. It was much easier to talk about my sister Betty giving birth in Japan to a healthy son named Timothy. His pictures already dominated the small family living room. Not wanting to go to bed too early, Sydney and I walked along the country road connecting our house with the nearby Indiana Dunes State Park. After only a few minutes, we were noticed by a patrol-car policeman suspicious of citizens outside their cars after dark. Sydney delighted in telling them that walking at night is taken for granted in England.

Two days later, fully rested, we set off for Pasadena, soon stopping off at the University of Illinois to let Sydney meet Salva Luria. After two hours of science over lunch with his lab group, we were back on corn-sided highways, feeling hot and sticky all the way to mid-Missouri. The 700 flat Kansas miles the next day left us zombie-like by the time we reached Colorado Springs, then somnolently filled with retired tourists who no longer had school-age children to look after. With Pike's Peak definitely no Matterhorn, we failed to be excited by the Rockies until we passed Gunnison. Then roadside signs told us that we were on the Million Dollar Highway and about to enter the "Switzerland" of America. We feared a local hoax but quickly found ourselves among spectacular jagged red sandstone peaks and the flaming fall-yellow colors of the aspens that dominated the foliage of the lower slopes. The high point of the day, however, was the charming 1890s towns of Silverton and Ouray. They still preserved feelings of the short-lived boom times that generated million-dollar silver yields.

After an overnight stay in Cortez, we drove on to see the massive, flat-topped buttes of Monument Valley, located where the states of Arizona, New Mexico, Colorado, and Utah all touch each other. Expecting almost impassable roads, we had stocked ourselves with much food and water but encountered no problems as we later drove past isolated groups of Navajo Indians. Ahead of schedule, we detoured up to the Grand Canyon that unexpectedly left us with neutral impressions. Perhaps it was because we were scenery-saturated or the canyon was so wide that the only way to get a true feeling for what it is like would have been to walk down to the bottom. But we wanted soon to be in Pasadena, so we were beyond Prescott at nightfall and by three the next afternoon, after a hot, unexciting desert drive, at Caltech.

First we went to the Athenaeum to find a room for Sydney, who planned a weeklong stay before going up to Berkeley to see Gunther Stent. My bags were also deposited there because Leslie and Alice Orgel were still living in my Del Mar Street flat and would continue to do so until they moved to Chicago at the New Year. Nervously, I made the brief walk to the Biology Building hoping to find a note from Christa. Initially my stomach sank when no such letter was apparent among the junk mail accumulated during my three-month absence. Suddenly I realized that the envelope bearing the imprint of the Gravity Research Foundation was, in fact, from Christa. Immeasurably relieved, I sought out Leslie to tell him that Sydney was at the Athenaeum. Never exuding praise for pedantic science, Leslie greeted me with genuine enthusiasm, stating he had been completely bored over the summer—it had had no redeeming feature of any sort. With Alice across Los Angeles that evening practicing medicine, Sydney, Leslie, and I were soon having supper along Colorado Avenue where we were not held back by the non-alcoholic limitations of the Athenaeum.

The next day Sydney was keen to learn about science while I drove to Colorado Avenue to find a haberdasher to make the first RNA Tie Club tie. Among my real mail was a letter from Geo Gamow containing a life-size paper pattern of his proposed design. With it in hand, I had no difficulty in finding a men's shop that promised to make ties for only $4 each, and the first one might be ready in a week.

Only a day and a half of undiluted Caltech science passed before Sydney's brain became exposed to more facts than he could politely handle. To get his courage up for one more day of American earnestness, we took Thursday afternoon off to experience Forest Lawn, the gaudy cemetery that Evelyn Waugh celebrated through *The Loved One,* his novelistic account of burial practices in Los Angeles. Alas, there was little for us to experience. Its unrestrained statuary—some sculpted to embrace the sale of final-destination plots and others to reflect calls from the Creator—in no way made up for our smarting eyes. The enveloping yellow smog cut off even the nearest mountains.

Although the phage-oriented lab of Max Delbrück and Renato Dulbecco's animal-virus group bustled with after-vacation vitality, I was most relieved not to be attached to Delbrück's group anymore. Instead I was on my own in a big office room on the floor above, just across from George Beadle's office. It provided the space for serious model-building now that Alex Rich was at the National Institutes of Health and Linus Pauling's chemistry department benches were no longer at my disposal. With Leslie leaving just after Christmas, though, I worried about my shortcomings in chemistry. In 1953, they might easily have cost Francis and me the double helix. If Jerry Donohue, the Caltech-grown chemist, had not had a sabbatical year at the Cavendish, we might not have been the first to appreciate the structural significance of Erwin Chargaff's base pairs. Learning some real chemistry now could only be for the good. So before the weekend started, I arranged with Linus's secretary for a Monday-morning appointment.

Fearful of exposing the true depth of my lack of chemistry, I anxiously entered his large office in the Crellin Laboratory. But Linus soon put me at ease and let us talk as if we were almost equals. Quickly we moved from DNA to RNA and my feeling that only after we had solved the RNA structure would we understand how RNA templates order amino acids during protein synthesis. The trouble was, our current X-ray pictures were too disordered to give us a reasonable chance of finding the helical parameters for successful model-building. Seeming more relaxed than I had ever seen him, Linus cautioned me about being in too much of a hurry for another big breakthrough. Instead he urged me to spend the coming year beefing up my knowledge of statistical mechanics and quantum mechanics. Unclear whether Linus had any idea how little physics birdwatchers usually learn, I rapidly seized his alternative suggestion that I attend his fall-term lecture course on the "Nature of the chemical bond." His first lecture was at 9 the next morning, and I left his office knowing that I would be there.

Crick had already written from England saying how wonderful Cambridge still was and offering me a place there should I want to go back. And, remembering the Gordon Conference, he wrote, "How is Christa?", a question that I feared might take longer to settle than the RNA structure itself. Here I took comfort that Margot Schutt, my shipboard Henry James girl from Vassar, still had me—at least slightly—on her mind. A new letter from her revealed that she was off the next day to teach history in what I took to be a posh girls' school in the fox-hunting region west of Washington.

With Sydney still around, there was no point for Leslie Orgel and me to get serious until he left. We also had Rosalind Franklin briefly on hand, a visit that we feared might easily go awkward. She had come down to Woods Hole late in August just after her arrival from England. Learning of her plans to go West, I asked her if she wanted to accompany Sydney and me to Pasadena. But with her six-week travel plans fixed, she declined, knowing she would see us at Caltech. To my relief, she now proved the opposite of unpleasant, eagerly telling me about her recent X-ray diffraction studies on tobacco mosaic virus (TMV). Happily, they both confirmed and greatly extended my 1952 Cambridge proposal that the protein subunits of TMV were helically arranged. Observing her current friendliness, Leslie took us aside to say that Rosalind's past reputation as difficult to get along with was unbelievable. In his mind, she had been judged most unfairly when working on DNA. Sydney and I had to agree, but still wondered whether our trip West would have been so lighthearted if Rosalind had been with us.

Before Rosalind went on to Berkeley, she and I had dinner with the Paulings in their expansive foothills ranch-style house. It was still largely surrounded by desert plants except for a well-manicured broad lawn beneath the glass-fronted living-room doors. Our invitation was last-minute, given during a brief, clear respite from the thick smog so characteristic of Pasadena's early fall. The 6000-foot San Gabriel Mountains above us stood out magically as Rosalind and I drove up from Caltech. Before arriving at the Paulings', I wondered what to say to Ava Helen. My concerns were needless as both she and Linus seemed pleased to learn more about me. In turn, I wanted to learn more about Linda, Peter's younger sister. She had just departed for Europe to be near Peter, much in the same way as my sister followed me to Europe in 1951.

Later, our conversation turned to Francis and his recent article in the October *Scientific American*. I had been jealous earlier when I saw that he, not I, had been asked to write about the double helix. But then my, not his, picture was in *Vogue* with its readership of stylish women, who had never read, much less heard of, *Scientific American*. In it, Francis's prose was clean and concise. Only in describing the base pairs did he let himself go, writing that he expected that an enthusiastic biologist would someday call his twins Adenine and Thymine. When Erwin Chargaff learned of Crick's exuberance, he agreed that this would probably happen but questioned whether enthusiastic was the right adjective.

Pasadena: October 1954

WITH SNOW SOON forecast for the High Sierra, Leslie Orgel and I used the first weekend in October 1954 to drive north to Lone Pine and from there to the pine tree–surrounded campground that lay 8000 feet below the summit of Mt. Whitney, the highest mountain in California. On our second night, neither of us slept well in our sleeping bags, partly due to the 12,000-foot altitude of our campsite but also because of some wretched bouillon that made our stomachs feel filled with lead. When a trace of morning lightness hit, we climbed upwards under cloudless conditions, finding only the last 200 feet difficult. Not enough sleep and too little oxygen had made us slightly dizzy. After quick looks from the flat, rocky, 14,496-foot top to the expansive scenery below, we retraced our steps towards more oxygen.

The 13-mile descent took less than four hours, and we were down to Lone Pine in time for a coffee-shop lunch. On the ride back, we talked about Rosalind Franklin's new X-ray data that questioned our simplistic picture of tobacco mosaic virus's RNA component forming a solid cylindrical core. From what Rosalind had just told us, combined with more convincing X-ray data that came later in a letter from Yale, it became clear that TMV's center was filled with water, not RNA. And because there was not enough RNA in a single TMV particle to form an external shell, the RNA chains within TMV particles must be tightly interdigitated into its helically arranged protein components (molecular weight 17,000).

Back at Pasadena, Leslie and I began making molecular models to explore whether it was possible chemically for intact double-stranded DNA molecules to serve as templates for the single-stranded RNA molecules. We wanted to see if single base pairs might function as template bits that specifically attract single RNA bases to generate base triplets. If so, we should be able to use our Pauling–Corey space-filling models to build a pretty triple helix consisting of a DNA double helix hydrogen-bonded to a single-stranded RNA molecule. Soon we focused on a scheme whereby adenine–thymine (A–T) base pairs attract adenine, T–A bases bond to uracil, guanine-cytosine (G–C) base pairs attach guanine, and C–G base pairs bond cytosine. Conceivably the single hydrogen bonds underlying the selection of specific RNA bases were strong enough for an accurate RNA-making scheme. No longer bored, we celebrated by walking several blocks after dinner to Lake Street for hot fudge sundaes. Afterwards, I penned a brief progress report to Christa Mayr, which I took the next morning to the post office. By then, to my great relief, her letters from Swarthmore ended "*I love you.*" And I replied likewise.

Already, I had given Dick Feynman our triplet scheme for making RNA but quickly sensed during subsequent lunchtime talk that he wanted solid facts, not dreams. So our conversation shifted to the Gravity Research Foundation, with which, to my surprise, Dick was well acquainted. It gave physicists license for far-out antigravity schemes without any of their friends thinking that they had gone off the deep end. Already some of Dick's friends had submitted prize essays. Quickly Dick spun out a proposal that he knew was no better than crap and could never dignify with his name. We thought it might be fun to see how it would fare under Leslie's and my attributions. Briefly anticipating its prize money, we soon learnt the same idea had already won an award several years before.

The prospect of yet another dead Pasadena weekend led me back to the Sierra by Friday evening, this time going with Renato Dulbecco and his adolescent son, Pietro. Hiking in from Mineral King, we were alone except for deer hunters, amazed that we were there only for walking. The crystal-clear air of our two nights on the high open meadows was in total contrast to the hellish thick smog that greeted us on our return to the Los Angeles basin. Making the smog even harder to take was the fact that I could not come up with an RNA backbone model that did not have impossibly close atomic contacts. With deep frustration, Leslie, a Swiss chemist friend, and I drove north one afternoon into the San Gabriel Mountains until, at 7000 feet, the smog vanished. There on a roadside table our Swiss friend read about quantum mechanics, while Leslie and I wrote an introduction to the triplet paper that we knew might never see the light of day.

The following morning I still saw no way to save our new scheme and our lunch was of unmitigated gloom. There seemingly was no way to get the RNA backbone atoms out of each other's way when they were confined to a radius of 7.5 angstroms. Only back alone in my office did I suddenly see a beautiful way out of our dilemma. If the nascent RNA chain was based on anhydride phosphate groups, formed by splitting a water molecule from each sugar–phosphate group, the resulting triple-esterified sugar–phosphate backbone formed a very compact RNA helix that fitted perfectly into the DNA double helix. The thought that RNA might be synthesized in the anhydride form, while radical, nonetheless was chemically possible because triplet-esterified organic compounds, though unstable, can be synthesized and studied. In fact, an inherent instability of the backbone, while bonded to its template, might be a great advantage. As soon as the cyclic bonds break, the hydrogen bonds holding the base triplets would also break and the complete RNA chains peel off their DNA templates.

Later, Leslie was predictably bowled over by the beauty of the structure. Like me, he saw the need to explain why the cyclic bonds always break to form $3'$–$5'$ bonds as opposed to $2'$–$5'$ bonds. But some $2'$–$3'$ cyclic nucleotides enzymatically break this way and so he was not overly concerned. Immensely relieved that my brain was still capable of a big leap, I no longer felt the need to leave Pasadena for the weekend. Suddenly I had much to do and decided not to join Leslie and Alice on their coming trip to the Sierra so that I would have the time to work out the coordinates of the anhydride backbone as well as finish the short manuscript that Leslie and I had rashly started on the smog-free Angeles Crest Highway the weekend before.

Tess and Victor Rothschild

Meanwhile, the first RNA tie had just been finished, and I unabashedly wore it to dinner that night at the Athenaeum. By then Victor Rothschild and his wife Tess had just arrived from England and were staying in the suite for important guests. A British member of the illustrious Rothschild banking family, Victor had trained as a zoologist and had come to Caltech for a month to do experiments with Caltech's sea-urchin specialist, Albert Tyler. In Cambridge, Victor had worked closely with Av Mitchison's brother, Murdoch, and it was in their Zoology lab that I first met him. Hopefully that night he and Tess would see the RNA Tie as a thinking brain's response to academic dullness. But, perhaps because of jet lag, they were not up to appearing and feigning pleasure that they had exchanged the coal-fire smell of English air for the automobile-generated yellow acridity of Pasadena.

By Sunday night I had a handwritten first draft of our manuscript, hoping to have it typed by the next afternoon so that Leslie and I could give Linus Pauling a copy after his morning lecture on chemical bonds.

George Gamow's original cutout-paper design for the RNA Tie Club tie

When he first read the manuscript, Linus warned us not to publish until we had more evidence. Like Dick Feynman, whom we saw later in the day, his first reaction was complete disbelief. But after we had fully explained our arguments, Linus slowly became more open-minded and thought publication might be worth the gamble. He saw the beauty of a template with an inherently unstable product that would automatically peel away from its generating surface. While we might be far out, he said, we might also be right. In particular, Linus recalled that several years before he had proposed an analogous argument about the transition-state nature of the intermediates in enzyme-catalyzed reactions. In spite of some opposition, Linus thought it was one of his best contributions to chemistry. After seeing Linus, however, I worried whether our anhydride RNA backbones really existed. Would any structure known either as OW (Orgel–Watson) or WO (Watson–Orgel) ever sound convincing?

For the rest of the week I talked much with Gunther Stent, then down from Berkeley, about his recent experiments using heavily ^{32}P-labelled bacteriophage particles to test whether the two strands of the double helix came apart during DNA duplications. Because his results were not easy to explain, Leslie and I spent Sunday heretically exploring the possibility that the two strands of the double helix never separate but instead serve as the template for two-stranded, base-paired daughter products. Underlying this heresy was the possibility of creating base quarters formed by specifically hydrogen bonding two base pairs to each other. The fact that we could not see how to build a perfectly regular four-chained DNA helix might, in fact, be a plus—its inherent instability would cause the daughter double helix to untwist away from its parental double helix. But our enthusiasm was soon dampened when we realized that Gunther's experiments were still far too preliminary to interpret.

The first real social life of the Caltech fall had come the day before when the Paulings gave a large tea party at their home to which 150 people were invited. My mind then did not easily lend itself to small talk and I had a long conversation instead with André and Marguerite Lwoff. They had just come from Paris to use new animal-virus plaque techniques being developed in Dulbecco's lab. Sensing my lack of enthusiasm for Pasadena, André suggested I spend a year with them at the Institut Pasteur. Later that evening, at a small square dance in the Delbrücks' living room, André asked why I didn't join in—I had been such a keen square dancer at Cold Spring Harbor the year before. Then he asked where my brown-haired muse of that year had gone, hinting that he knew why I was not in a mood to dance.

I continued faithfully to get to Linus's early-morning lectures, but was too excited by our RNA thoughts to spend any noticeable time reading the outside class assignments. So with each subsequent lecture I found myself understanding less and less. Even more incomprehensible was the public lecture Dick Feynman gave on quantum mechanics and whether physical phenomena are chance events. Given Dick's irresistible personality, the lecture was total fun, given in the Physics Department to a packed audience that applauded before he even began speaking—respect that no other lecturer at Caltech then commanded. Physics also entered my life with a cheerful communication from George Gamow, excited by his apparent finding of a financially rich backer for the RNA Tie Club. The U.S. Army Quartermaster Corps would not have been my sponsor of choice, but their promise to fund biannual gatherings of the club was not to be sneezed at.

One Friday morning late in October, rumors began to float that Linus had been awarded the 1954 Nobel Prize for Chemistry. Such reports often proved false for others, but by the afternoon he was receiving official congratulations and I got word from his secretary of a celebratory cocktail party at his home. When I was in England, at Cambridge, Peter Pauling had told me that the annual October awarding of the prizes had become a source of much tension in the Pauling household. It had been more than 20 years since Linus had used quantum mechanics to provide fundamental insights about the nature of the chemical bond. Particularly seminal had been his 1931 breakthrough in understanding the tetrahedral manner in which carbon atoms form chemical bonds. Although he had received many other prizes, the failure of the Swedish Academy to give him due recognition was increasingly a source of deep rankle.

But no sense of past hurt was evident at the party the Paulings gave the next night: champagne flowed copiously. Linus and Ava Helen had already decided to fly over the North Pole to Copenhagen and then on to Stockholm. All their family would join them in Sweden, and then Linus and Ava Helen would afterwards go on alone around the world, stopping in India and Japan for Linus to give lectures. Most of the guests were Linus's age, and I felt myself out of place by being 10 years younger than any of the other celebrants. Victor Rothschild was there alone, as Tess had flown back to England to be with their children, and I had somebody to gossip with. Later, the Lwoffs and I speculated on how Linus's life might be altered by the prize.

I had thought the occasion would have been one where Ava Helen was surrounded by the wives of Linus's long-time chemistry department colleagues. But they did not flutter about her, and we had the time to speak about Peter and the danger of his going too long without a firm direction in his research. Sensing that I had not yet found myself socially, Ava Helen suddenly revealed that she had long found herself bored by the heaviness of Caltech social life in contrast to the liveliness of more political occasions across Los Angeles in Hollywood. Soon after, she moved on to a nearby Caltech couple and our frankness, possibly too close to home for comfort, was lost in the ample champagne that remained available long into the night.

Pasadena and Berkeley: November–December 1954

WITH MY HEAD still heavy from not enough sleep and too much alcohol, Manny Delbrück, Renato Dulbecco, and I early the next morning drove east along the base of the San Gabriel Mountains to the foot of Mt. Baldy. We wanted to climb it before winter snows converted its upper slopes into a site for skiing. At its 10,000-foot summit, the bright sun and sweaters kept us from feeling too cold as we peeled oranges and ate cheese sandwiches. I tried not to think how much more fun I would be having if Christa were with me. Nor was this the occasion to dwell on my relief at having just received a letter from the Harvard Biology Department, inviting me to give a job-seeking seminar. Already I had written back that I would go to Cambridge after I had visited my parents at Christmas. Leslie and I thus had a pressing deadline to clarify our ideas on how protein synthesis occurred. Even if our new model for how RNA is made upon double-stranded DNA templates proved to be correct, the much bigger question remained of how RNA itself functions in amino acid ordering.

The week that followed was initially dominated by the removal of a lower wisdom tooth. I masked the resulting discomfort by browsing through the mammoth 900 pages of *In the Matter of J. Robert Oppenheimer,* the verbatim transcript of the Atomic Energy Commission (AEC) hearings that had just led to the removal of Oppenheimer's security clearance. Before leading the Los Alamos team, which built the first atomic bomb, Oppenheimer taught both at Caltech and Berkeley and still had many Caltech friends. The hearing centered on whether Oppenheimer's reluctance to work on much more powerful hydrogen bombs reflected a hidden pro-Soviet agenda. In the 1930s Oppenheimer's politics were far to the left, and the question arose whether several brief meetings with American communist friends after he went to Los Alamos had transferred vital information to the Soviets about our two-billion-dollar Atom Bomb Project. Although many of his fellow physicists strongly testified for Oppenheimer's loyalty, there were several others—in particular, George Gamow's close friend Edward Teller—who counter-argued that Oppenheimer's behavior, in opposing trying to build a hydrogen bomb, had the smell of a communist. In contrast, Hans Bethe, also close to Gamow, testified strongly for Oppenheimer, stating that "the super" could so easily wipe out modern civilization.

The pervasive unease felt at Caltech with the Oppenheimer verdict was subsequently heightened by our learning from George Beadle that two Caltech scientists had recently been denied non-classified research grants from the National Institutes of Health (NIH) because of putative communist leanings. Deeply disturbed by political loyalty requirements, George thought Caltech must make a public stand and should possibly refuse more NIH grants until the recent decision made by the Secretary of Health and Welfare, Oveta Culp Hobby, was reversed. The first Caltech grant application to run into trouble, not surprisingly, was from Linus Pauling. His highly visible opposition to nuclear weapons, and appearances at so-called Communist Front Peace Rallies, was a deepening thorn in the sides of the many deeply pro-military Caltech trustees. In contrast, until Biology professor Henry Boorsok's grant ran into trouble, no one thought of him in political terms.

Before this political cancer spread further, George ("Beets") wanted his faculty's concurrence in letting Mrs. Hobby know the shortsightedness of her position. It too easily brought to mind political reliability tests imposed on Nazi-era German professors. Linus, perhaps feeling the government's distrust of him would never end, had his grant resubmitted under the name of his co-worker, Robert Corey. As such, it was soon funded. Henry Boorsok, however, did not have this way to stay scientifically alive. Fortunately Secretary Hobby soon reversed her decision, and the political litmus test vanished from the awarding of NIH grants. Here Beets's forceful intervention was helped by the publicly announced decision of the greatly respected Harvard protein chemist, John Edsall, to give back his NIH monies if Boorsok's grant was not funded.

Over the next several November weeks, Leslie and I returned to how RNA molecules form cavities specific for the side groups of the 20 different amino acids. None of the potential helical fold we considered for single-stranded RNA molecules generated even half-good holes into which the specific side groups would bind. Even subsequent daily visits of Dick Feynman to our model-building table did nothing to break our mental road jam. After an hour or so of looking at potential amino acid–binding surfaces, he invariably gave up and returned to meson theory frustrations. The week before, Dick and I had received essentially the same letter from a California rabbi asking our views on religious revelation and spiritual guidance. As an escapee from the Catholic religion, I wrote back that I had no interest in religion. But Dick wrote back a much stronger response, believing that calling crap, crap was the Brooklyn way to make your mark on this world.

Gamow was by then the possessor of the sole RNA tie so far made. Earlier in the week, he sent me an urgent telegram asking me to dispatch it to Washington by special delivery. He wanted to wear it to a coming weekend meeting of the National Academy of Sciences, where he was going to speak about his code-finding efforts. There he would pass the tie on to Melvin Calvin, the Berkeley chemist who was to talk at NIH several days later.

By Thursday evening, feeling mentally stale, I came into the Athenaeum expecting to eat quickly but noticed Victor Rothschild eating alone. He explained that much too often he found himself next to impassive souls interested only in their work, college football, and new cars. With Tess back in England, he was spending most of his days and nights at the lab. This evening, however, he wanted out from Caltech life. So we drove the Pasadena Freeway into the center of Los Angeles and then up Wilshire Boulevard to see the new Italian film *Bread, Desire, and Dreams*. Gina Lollobrigida starred. Our fantasies took off and we came back to Pasadena with much-improved morales.

The following Tuesday night, Leslie, Victor, and I had a very alcoholic occasion at The Stuffed Shirt. This was, in effect, a farewell dinner for Victor. To my surprise, he started talking about his British philosopher friend, Stuart Hampshire, with whom Tess had had dinner in New York on her way back to London. That morning Christa's latest letter had told me of listening to, but not understanding, Hampshire at a Swarthmore lecture. Over cognac at Leslie's flat, we talked more about Cambridge. To his delight, Leslie had just received one of its lectureships in Chemistry. This was totally unexpected because only rarely did Cambridge offer positions to Oxford graduates, and vice versa. In going back to the Athenaeum, Victor encouraged me to get back to Cambridge as soon as possible. Later, telling Beets of Victor's advice, I found him sympathetic to my wish to be near Francis again.

Several letters a week from Geo Gamow were then barraging me, one excitedly telling me of help from the Los Alamos computer whiz Nic Metropolis. Through him the powerful bomb-making Maniac Electronic Computer was also working on the genetic code. At the same time, Geo's life at home with Rho had become hellish, and he was moving temporarily into the Cosmos Club in Washington. To divert himself, he had started circulating a chain letter to the now-17 members of the RNA Tie Club asking them to choose an amino acid whose abbreviation would be inscribed on their respective RNA tie pins. Geo picked Alanine so he would be ALA, leading me to choose Proline so that I could be PRO. With the design already in hand, the tie pin could soon be made in Washington, giving Geo the opportunity to trap new acquaintances into asking why he had misleading initials all over his tie. Later he was the victim of his own joke when a hotel cashier in Chicago refused to honor his check, noticing that ALA did not correspond to George Gamow.

More seriously, Geo kept trying to find support for his original "diamond" code through examination of the newly available 165 amino acid sequences of the polypeptide hormone ACTH. Frustratingly, two double-letter sequences, Lys–Lys–Arg–Arg, ruled out his beloved diamonds. So he had moved on to a crazy triangular code that would simultaneously generate two polypeptide chains from a single DNA sequence. In this way, he hoped to explain similarities between the A and B chains of insulin. This was a mad idea that died almost as soon as it was born. Geo went on to construct empirical curves relating the number of different amino acid neighbors found to the total number of already sequenced amino acids. Hopefully, the new calculations coming off Maniac would point to restrictions in the number of neighbors that any given amino acid possessed. If no restrictions existed, then completely different sets of base pairs must be used to code for successive amino acids along the polypeptide. Geo disliked this possibility because, if true, there would be no way to guess the nature of the genetic code through examining amino acid sequences—the only tool at his disposal.

Geo also remained bothered by the DNA→RNA→protein relation and soon posted me a cartoon message asking why cells with DNA rich in AT base pairs contained RNA that had predominantly G and C bases. But this fact, which was already well known, only mildly disturbed me. I could explain it by postulating that genes rich in G and C are more commonly expressed than genes rich in A and T bases. Why this is so, I didn't know. But I saw no reason to doubt that RNA molecules are the templates that order amino acids during protein synthesis.

After seeing Victor Rothschild—now suitably dressed in a dark pin-striped suit for his return to "civilization"—off to the airport, Leslie came up with a bizarre idea that double-stranded DNA, not RNA, had the surface for attracting amino acid side groups. By slipping the two DNA chains past each and holding them together, not by hydrogen bonds but by divalent ion bridges, Leslie thought that amino acid side groups might be bound. But suspecting it was an extremely long shot, Leslie's brainstorm totally left my mind during an Italian meal cooked for André and Marguerite Lwoff on Saturday night by Renato Dulbecco's wife, Enucia. The next day the Lwoffs and I walked through the expansive Italianate grounds of the nearby Huntington Library, peeking into the building itself to see its two main artistic treasures—Thomas Gainsborough's exquisite *Blue Boy* and *Pinkie*. Smiling broadly, André admitted they were good enough for the Louvre.

Three days later, the warmth of southern California vanished as Leslie and I crossed over the Tejon Pass and descended into the foggy cold of the Central Valley. We were on our way to a Thanksgiving feast with Gunther and Inga Stent in Berkeley. The next day we went to the Virus Lab, hoping that Robley Williams's electron microscope would give vital clues about the shape of his TMV RNA molecules as well as some animal-cell RNA I had brought from Caltech. But both our samples gave rise to flat puddlelike masses, and we left no wiser than when we arrived. That evening, we predictably did not find gaiety in what Gunther enthusiastically proclaimed as San Francisco's most existentialist night club.

Upon arriving back at Caltech, I faced the fact that in only a month I must give my Harvard job seminar. I was uncertain about what new ideas I could talk about. With each passing day, Leslie and I got colder feet about our phosphoanhydride RNA model that the month before had seemed so perfect. Although Leslie seldom doubted his acumen as a theoretical inorganic chemist, he had little feel for organic chemistry and its biochemical offshoots. There was no way to predict whether our DNA→RNA scheme had any chance of surviving expert scrutiny. In particular, the fact that we were using only one hydrogen bond to attract an RNA base to its corresponding DNA base pair was bound to raise questions of whether our scheme had the requisite accuracy needed for RNA synthesis. We were resigned to delaying submission of our completed manuscript to *Nature* until we had solid evidence. We had no clue about how to move ahead experimentally and my frustration was compounded further by our continuing inability to twist an RNA chain into any shape with true template properties.

At best I was in a restrained mood when I went to the big Friday night Athenaeum faculty banquet held to honor Linus just before he and Ava Helen flew off to Sweden. When he and I briefly chatted the day before, his very tweedy suit was perfect for the now almost cool weather, and he was very pleased with his circumstances. Highlighting the banquet was a skit satirizing Linus entitled "The Road to Stockholm." Its clever lyrics put everyone in a light mood, especially Linus, who announced this to be the happiest evening of his life.

Dick Feynman and I sat next to each other. Although we could not say it to others, we felt we might be Caltech's most obvious candidates for future Nobel awards. More privately the next afternoon, Dick told me that his contributions to physics were still insignificant compared with the great minds from the Bohr-Heisenberg era. Although now no theoretician outshone him, he wanted to come up with a more profound contribution to physics before he went to Stockholm. In the same way I felt the need to have more than the double helix below my belt before winning the prize. I did not want to be overpraised for what was not very difficult science.

The next evening I drove up to the Altadena home of Stuart Harrison, a British-born physician with many Caltech friends. In the mid-1930s, he helped establish the student health service but now he specialized in radiology and could afford his smart ranch-style house. Some months before at Mariette Robertson's house we had fun talking, and at the Pauling banquet he said I must come up soon to his and his wife's foothills' home. Finding they had no plans for the next night I asked "Why not then?" and, to my relief, was invited. I needed desperately to talk to a real physician. Persistent anxieties about Christa, heightened by the tone and decreasing frequency of her letters, were affecting my ability to concentrate either on model-building or preparing for my impending talk at Harvard. Falling asleep was becoming a nightly struggle, and without medical help I feared a nervous breakdown that would not only foreclose any offer from Harvard but also give Christa the message that I was emotionally fragile.

Spotting my unease upon my arrival, Stuart gave me two Scotch and sodas. Soon I no longer felt terminal anxiety and was being reassured that my plight was not unusual. He told me about the mid-thirties Caltech and its bohemian underlife that moved to a very different drummer than that governing the self-assured solemnity of its founding president, Robert Millikan. Many couples who then knew each other too well had cause for much anxiety. In particular, Stuart related the turmoil that accompanied the affair of his left-leaning first wife, Kitty, with Robert Oppenheimer, whom she subsequently married just before Oppenheimer began leading the Los Alamos atomic-bomb lab.

With my nerves no longer jangling, I returned to the Harrisons' home a week later for a dinner honoring a couple from Italy. A cheerful letter earlier in the week from Christa had put me back on the sleep track. No hint came through that anyone else was in her life, and she was looking forward to seeing me after Christmas. No longer in panic, I stopped intellectually freezing when looking at my RNA model and came up with a new single-chain RNA helix that repeated every 12 angstroms. It had the phosphate groups on the outside and the bases stacked next to each other 45 degrees to the perpendicular position. Tentatively this model had the potential for at least several good amino-acid binding holes. But with the Orgels off camping in Death Valley, the weekend passed before Leslie had a chance of seeing it. Later we realized that there was not enough time before I left for Christmas to see if the model actually had template features. Later I enjoyed Geo's quickly penned response comparing my potential template holes to a tiger's mouth.

Leslie and I, however, were still model-gazing when a thick letter arrived from Francis Crick containing a draft of the article on plant-virus structure that he and I had long planned to write. Full of flamboyant Crickisms, I saw the need to write them out of the final manuscript. Meanwhile, I was much enjoying words that Mariette Robertson sent me from Paris, where she now was living with her parents. Earlier in the fall she had spent several weeks travelling about the south of France with Linda Pauling—Linus and Ava Helen's daughter. More recently, Linda couldn't understand why Mariette had not yet congratulated her and Peter on their papa's prize. But Linus's prize had been for his chemistry, and Mariette saw no reason to see Linda on a pedestal whose base was the attractive personality that Linus had passed on to her and Peter. Reminding me that I was always warning her against the wiles of charm, Mariette revealed her concern that someday I might fall under Linda's blue-eyed blond spell. But she noted that during their recent travels the European students who flocked around Linda were not interested in her, and vice versa. So she thought I was unlikely later to fall for Linda. That night I easily fell asleep not that certain.

18

Northern Indiana, Cambridge (Mass.), and Washington D.C.: December 1954–January 1955

AN OFFER FROM Seymour Benzer to give a talk at Purdue University let me practice my Harvard job seminar the day after I flew to Chicago for Christmas. I drove my parents' car—the first they had had since the depths of the depression—to Lafayette. My journey took 90 minutes, most of them spent anticipating Christa, whom I would see in less than a week. Before my talk, Seymour happily told me more details of his elegant new genetic tricks for determining the location of mutational changes within a phage T4 gene. Utilizing the observation that $T4r2$ mutants do not grow in an *E. coli* strain containing phage λ, over the past year he had demonstrated the strictly linear order of several hundred different $r2$ mutations, each of which he believed represented base-pair changes along DNA molecules.

Invariably, Seymour worked late into the night and seldom arrived at his lab before lunchtime. But the day I arrived Seymour was there by 11:30 a.m. to take me home for a lunch prepared by his always upbeat, diminutive wife, Dottie. As it was the last day of classes before the Christmas recess, Seymour warned me that my audience might be thin. So I was pleasantly surprised by a totally filled lecture room. The moment I started speaking, I became excited again by the beauty of the double helix and how its complementary base sequences should provide the structural basis for DNA replication. I went on to argue that RNA must be the information-bearing molecule that carries genetic information from the chromosomal DNA to cytoplasmic sites of protein synthesis.

For most of my lecture, I talked about how single-stranded RNA chains might be made on DNA templates as well as how they in turn might serve as the templates for ordering the amino acids along polypeptide chains. Then I made it more than clear that no experimental support existed as yet for any aspect of our RNA triplet model: single RNA chains might well be formed upon single-stranded DNA templates using the base-pairing rules of double-helical DNA. Almost in passing, I mentioned that base quartets could be formed in which two base pairs related by a parallel diad are held together by two hydrogen bonds. When thinking about how DNA replicates, I said, we should not automatically rule out intact DNA double helices serving as templates for a second double helix. On the other hand, the resulting quadruple helix was not a regular structure with its exact configuration a function of its underlying DNA sequences. I ended my talk by emphasizing the positive role model-building would have in investigating such alternative template mechanisms. When the questioning started, I worried that some true chemist would pronounce my paper chemistry schemes not worthy of public presentation. No one in the audience, however, proved the slightest bit unpleasant, and I drove back to my parents' home feeling upbeat about my forthcoming Harvard debut.

Christmas Eve was my mother's fifty-fifth birthday, and for supper we had the oyster stew that always marked the occasion. With my grandmother Nana now dead and my sister Betty married and in Japan, the evening was muted. My parents knew that I was apprehensive about my forthcoming talk and whether or not Christa was yet ready for marriage. Before our Christmas meal, dominated by turkey with dressing and cranberry sauce, Dad and I took off for a brisk walk to Lake Michigan, hoping to get a respectable scorecard for our Christmas bird census. That Christmas day, however, the birds were sparse and only the sounds of juncos, chickadees, song sparrows, and the occasional cardinal caused us temporarily to halt our determined movements against the frigid winds coming off the lake.

Two days later in Boston, the Mayrs met my plane and took me to their Washington Avenue flat. The following night they hosted a small dinner party for Boris and Harriet Ephrussi, who next day were going down to New York before taking the boat back to France. Boris was his usual frank self in assessing their several-month stay among the Harvard biologists. The corn geneticist, Paul Mangelsdorf, was about to retire and Boris had wondered whether he was being looked over as his potential successor. Complicating this expectation was the question of whether Paul Levine, then teaching genetics, would be promoted to tenure despite his, up to then, unexceptional research on *Drosophila*. Boris, however, did not feel he would aid his own cause by telling Harvard that it might profit if Paul went elsewhere.

The next day the Mayrs were keen to see their New Hampshire farm in winter white, and we drove up for a day visit, knowing that the heat of its wooden stove would eventually warm up the kitchen. Soon after our arrival, Christa and I made a long circular walk that gave us the isolation that I had so long anticipated. As soon as we were out of sight, Christa and I started kissing and later hugging, lying on the powdered snow. When we arrived back to a now-warmed up kitchen, I sensed I had been too bold on the snow, and regretted it. Two days later, when finally again alone on a bus taking us to the Szent-Györgyis' Hungarian New Year's Eve party, we talked about everything except ourselves. Even at the stroke of midnight, when we had finished Albert's lovingly cooked suckling pig, Christa only wanted to hold hands with me. But on the way back to Boston, she expressed eagerness to come down a week from then from Swarthmore to Washington. There I would be spending my last weekend East with Alex and Jane Rich.

With Christa about to return to college, I moved into Dana Palmer House, the splendid Federal residence where Harvard puts up official visitors. As soon as I signed the guest register, Paul Doty leafed back to the page where Winston Churchill's signature stood out. Then he pointed to the names of persons whose stays there meant they were destined for Harvard offers. At lunch the next noon at the Faculty Club, I rose to the challenge of the horse-meat steak that had remained on the menu now 10 years after the end of the war with Germany and Japan.

For the next several days I was paraded through the offices of most of the Biology faculty, staying in good form almost until the time of my 5 p.m. talk. Increasingly anxious over whether I was going to spout more than hot air, I spoke in a low voice that only occasionally made it to the back of the lecture hall. Happily, the senior biology professors heard all my words because, aware of the wretched acoustics, they invariably sat in the front rows. The lecture hall was a bad afterthought of a 1934 building that the General Education Board of the Rockefeller Foundation funded to promote important research in biology. Depression-gripped Harvard failed its donors' expectations to put more of its own monies towards the building, leaving the entire north wing unfinished.

That molecules were still an unimportant feature of the Harvard biology scene made my job-seeking task easier. Virtually none of the faculty had the background to know whether my thoughts about RNA made sense. Only the several true chemists in the audience had the training to spot way-out nonsense. But neither Konrad Block or Frank Westheimer, both newly recruited from the University of Chicago, embarrassed me through questions that implied that my Caltech efforts of the past four months were totally speculative. So I could truly relax during the well-mannered, slightly boring dinner that followed in the Faculty Club. There the possibility of my wanting to leave Caltech was never brought up. I slept well that night with the Dotys reassuring me over breakfast that my talk was sensational.

The train got me to Washington the day before Christa was to arrive, giving me time to let off my romantic misgivings to Jane and Alex Rich at their small rented house near the Chesapeake Canal. Jane said that Christa and I would enjoy Georgetown and, after I met her train, we had a Wisconsin Avenue coffee-shop lunch. There we talked about her increasing yearning for a junior year at the University of Munich, where she could become fluent in her parents' native language. She did not want to follow in the footsteps of her mother's early life. The early twenties for Ernst had been one of romantic adventure, collecting birds in the Solomon Islands off New Guinea, whereas Gretel's equivalent years involved caring for two infant daughters in a foreign land on a depression-level, museum-curator's salary. With each new sentence I feared Christa would blurt out that we should cool our romance, not seeing each other again until after her European experience. But at the end of lunch she clearly expressed a wish to see me abroad. Particularly, she wanted to see if the old Cambridge was overwhelmingly more beautiful than its New World equivalent.

By the time the Cabin John trolley had let us off near the Riches' house, Geo Gamow had already arrived. For the previous two weeks, he had been lecturing in Florida, and later taking in more sun at St. Augustine. Now he and Alex were mulling over the long article they were writing on the "Coding aspects of information from nucleic acids to proteins." It was to appear as a solicited review and, although speculative, most of their arguments were original and unlikely to be criticized by a reviewer. Already half-finished, it contained many of the conjectures from the Woods Hole coding week as well as analyses made possible by more recently published amino acid sequences. There was no doubt that Geo's diamond code was now dead as was any part of the triangular code that could be precisely formulated.

Moreover, the fact that closely related forms of insulin—from the cow, pig, and human—differed from each other by only single amino acid replacements had to rule out any overlapping code in which single base pairs in DNA helped determine more than single amino acids. Unpleasant as this conclusion was to Geo, it was supported by the data coming from the Los Alamos Maniac computer. It was telling Geo that the still-limited number of observed amino acid neighbors was that expected from a completely random distribution of amino acids in polypeptides. Very likely each amino acid was determined by its own set of non-overlapping bases, all specific permutations of the four bases A, G, T, and C taken three at a time (e.g. AAA, AAG, AAC, etc.).

The conversation over dinner turned much less serious. We were all curious to know what Crick was now up to. John Kendrew had written me in November that I was badly needed at the Cavendish to keep Francis in order. If his intellect could be focused on RNA, he might have less time for orations. Now, John reported, Francis consumed whole days talking furiously about the secrets of life and how he would become famous by writing bestsellers. Clearly he was still annoyed by my decision last year to keep the double helix off the BBC. Apparently Francis now talked of a broad Reith-like radio series on "the meaning of life."

That evening we all went on to a party given by Dick Roberts, a physicist at the Carnegie Institute of Washington. Although he had done important neutron experiments that helped start the atomic-bomb project, he arrived for the 1948 Cold Spring Harbor Phage Course with golf clubs and bag in his car's trunk. Moreover, he became interested in extrasensory perception, giving an evening seminar attempting to show that he could guess the faces of unturned playing cards. Tonight, on home turf, he was less loony and wanted to use isotopes to study the metabolic stability of DNA and RNA. As expected, Geo tried to bring life to the party through his card magic, but it fell a bit flat as too many of the guests had experienced it before. Those not in the know assumed that Christa and I were a real couple, but the hollow feeling in my stomach told me otherwise. One other fetching face might have made the evening semi-passable, but this was an occasion when alcohol, not beauty, was the tool to overcome academic frustrations.

As I put Christa on the train back to Swarthmore, she promised to write as soon as she got gossip from her parents about Harvard's reaction to me. The next afternoon I gave a lecture at NIH before a set of impassive, question-less biochemists and the following day was on the plane back to Los Angeles. In boarding it I regretted not arranging that day to go out to Foxcroft, the posh girls' boarding school located in the Middlebury horse country to the west of Washington. There Margot Schutt, the evasive Vassar girl who held my eyes on the boat back from England, was now teaching history. In reaction to Foxcroft's reputed horsiness, I guessed she might be in need of some human warmth.

19

Pasadena and Berkeley: February–March 1955

BACK IN PASADENA I had to face up to the loss of Leslie and Alice Orgel. At the new year, Leslie moved on from Linus Pauling's group to be at the University of Chicago with Robert Mulliken, whose theoretical ideas about chemical bonds were in the ascendancy. I was effectively left with no one to speak unreservedly to about matters of the heart and also, with only one exception, Don Caspar, about RNA.

Don had just arrived from Yale formally to be a postdoctoral fellow with Max Delbrück. For his Ph.D. thesis, Don had used X-ray diffraction techniques to establish the hole in the center of long, thin, pencil-shaped, RNA-containing tobacco mosaic virus (TMV). He was keen to do further X-ray diffraction studies on plant viruses, but found the Caltech crystallographic facilities not up to the task. Together we began using the Spinco analytical centrifuge belonging to the plant physiologists. With this instrument we could measure the size of TMV RNA molecules that Norman Simmons from the University of California at Los Angeles was preparing for me. When purified, this RNA might give much better diffraction patterns than those used the year before by Alex Rich and me. Don was always around for an evening meal because he had no local girlfriend and our conversations were more fact-laden than relaxing. Usually we avoided Athenaeum suppers and became regulars at a Van de Kamp restaurant, east of Caltech.

My morale was not helped by seeing the just-released Japanese film *Gates of Hell,* about a samurai warrior going insane through his love of a married woman. Everyone in the departing audience looked emotionally stretched as we walked towards our cars in the blowing rain, and I was relieved that I need not return to the ghostly blah of an Athenaeum room. Instead I headed towards my small, reclaimed flat that the Orgels had occupied since my departure last June for Woods Hole. Gracing it now was an engraving done in 1840 of an Oxford college chapel from Victor and Tess Rothschild that served as their seasonal holiday card this year. Also comforting me was Geo Gamow's recent special-delivery message to Alex and Jane Rich, noting that he "highly approved of Jim's juvenile girl in tow."

Only several more days went by before twin sources of anxiety temporarily abated. A letter from Christa finally arrived, letting my stomach no longer seem to press against my legs. In it, she wrote that a Harvard appointment prospect looked excellent though the Biology Department would take several months to act. Ernst was working to convince his department that I was at heart a biologist, not a biochemist, and to that end kept referring to my deep interest in birds. No matter what they decided, however, I knew I was getting nowhere in Pasadena and should leave before Caltech came to the same conclusion. Already in the mid-fall I had written the National Science Foundation (NSF) in Washington to see if they could provide me senior fellowship funds to let me work with Francis again. Happily they said this might be possible, and I was soon sending them a formal proposal that would give me a yearly salary equivalent to that Caltech then provided.

George Beadle knew that my girlfriend situation was driving me bananas and took my decision to leave with a regretful smile. Beets, nonetheless, kept encouraging me to fall for some local girl, and for a brief January moment I had hopes that a pretty small brunette from a Kansas academic family might be fun to be with. But our dinner date never got animated, and soon she was regularly having coffee with a young physicist. The only social events that had chances of coming alive for me involved academic visitors. Much fun came when I drove with Denys Wilkinson, the Cambridge nuclear physicist and avid bird-watcher, across the Mexican border into Baja California, whose coast-line below Tijuana sparkled bright blue. Early in the afternoon, sensing we might be close to the ocean, we stopped below Ensenada near the small adobe-filled village of San Simon and hiked westwards to the coast. Unfortunately the tide was out with most shorebirds too far away. The following morning, we panicked when my car's battery did not turn over. We were forced to walk several miles for help from two Mexican farm laborers just able to push our car fast enough to start the battery. Immensely relieved—Denys was expected that night for a Pasadena faculty supper—we recklessly risked axles and tires going fast over the 100 rutted-gravel miles that lay before the paved roads that would finally get us back to Caltech.

At the end of January, Geo Gamow descended again on Caltech. His efforts to get Quartermaster Corps money for RNA Tie Club meetings had collapsed, but now he thought that the NSF might fund a small elite RNA and Protein Synthesis meeting in Boston in mid-June. I helped him prepare a list of names that he could send on to Washington. While the Caltech biologists at first welcomed Geo's latest visit, the departure after four days of his large frame was a relief even to Don and me. Some four quarts of liquor store whisky were consumed, not counting those drunk at Stuart Harrison's Saturday-night dinner party.

When I discussed the Robert Oppenheimer affair with Geo earlier that day, I felt let down when I learnt that Geo felt Edward Teller had the right to say what he believed about Oppenheimer. But Geo thought that Teller behaved badly in hogging credit for the hydrogen bomb when the key intellectual trick came from the Polish-born mathematician Stan Ulam. When I protested that Oppenheimer had not been fairly treated, all Geo could reply was that politics was dirty and nothing could be done differently. Late Saturday night, while I was driving him back to the Athenaeum, Geo's arrogance had gone, and he told me that life had its difficult transitions. He did not mention his marital troubles with Rho, but there was little doubt they were bothering him.

On the way to Santa Monica, where I drove Geo for a Rand Corporation conference on interplanetary travel, we talked about why viruses—such as tobacco mosaic virus (TMV) and poliomyelitis virus—contained RNA and no DNA. Did this mean RNA also functioned as a master genetic molecule capable of exact self-replication? If so, were there two forms of RNA or could one molecule be both the gene and the template for protein synthesis? This might be why RNA from TMV and the ribosomal particle sites of protein synthesis gave identical X-ray diffraction diagrams. In the car, Geo's happily still Scotch-free responses were much to the point, again revealing the super mind that so early in his life catapulted him into the ranks of the very best physicists.

From Santa Monica, it was less than a half-hour drive to the University of California LA lab of Norman Simon, from whose lab I got a new RNA preparation. Initially I got overexcited when its high negative birefringence implied bases oriented strongly perpendicular to the putative helical axis. The X-ray patterns these fibers generated, however, showed no better orientation than those Alex Rich obtained the year before. Increasingly I wanted to find out what TMV RNA molecules looked like in the electron microscope. Last November, while in Berkeley, my attempts with Robley Williams to look at TMV RNA had revealed only fibrous contaminants present in the city tap water. By using distilled water as well as much better RNA, future answers might be there to get.

Before driving up again to the verdant Berkeley campus, I had a brief anxiety attack over a letter from George Wald. Over the holidays he had told me that he would write when something happened at Harvard, and so I hesitated to open his letter. But its purpose was only to ask me a simple question about DNA to which was appended the sentence "the affair Watson is going well and be patient." The next afternoon, the Physics Department auditorium was jammed for Dick Feynman's report of a recent East Coast meeting on the many, too many fundamental particles being discovered. Their inherent complexity made our RNA paradoxes mere children's games as opposed to the adult dilemmas that so frustrated Dick, who wanted a grand scheme for nuclear forces like that accomplished for the atom in 1926.

At Berkeley, Don Caspar and I were to give a joint seminar on TMV for which Wendell Stanley would pay us $50. Instead of barreling straight up Highway 99, we cut west to Taft and got caught in a speed trap for doing 65 miles per hour in the middle of nowhere. The unexpected beauty that came with crossing the Gabilan Mountain range into the Pinnacles National Monument happily made the stupid fine worthwhile. The next four days of Berkeley spring, punctuated by gentle English-like rain showers, kept me in high spirits despite our failure

Columbia University

College of Physicians and Surgeons

630 WEST 168TH STREET

NEW YORK 32, N. Y.

DEPARTMENT OF BIOCHEMISTRY

CELL CHEMISTRY LABORATORY   January 3, 1955   OFFICE: ROOM 12-201

TEL. WAdsworth 3-2500

EXT. 7479

Dr. J. D. Watson
Biology Department
California Institute of Technology
Pasadena, Calif.

Dear Doctor Watson:

At the urgent behest of Gamow I am writing to say that I have
completed the complicated requirements of membership in the "RNA Club."
This, I am afraid, entitles me to a necktie.

With best wishes,

sincerely yours,

Erwin Chargaff:eb

again to see genuine RNA fibers in the Virus Lab's electron microscope. On the way back, I felt confident enough of my driving to risk the precipitous voids below the coast highway that we followed after spending the night at Big Sur.

The dejection, which inevitably hit me on my return to Pasadena, was not helped by my need to give, two nights later, a pepless Sigma Xi lecture before a general audience that included physicists and chemists as well as biologists. Speaking about DNA alone would have depressed me, so I ended my talk with the dilemmas coming out of finding viruses with RNA as their genetic material. It was key to find out whether each TMV particle contained just one RNA molecule, whose length, in turn, defined the 2800-angstrom length of infectious TMV particles. Unfortunately the measured molecular weights of TMV RNA suggested each TMV particle might contain some 10 to 20 separate RNA molecules. If so, I wondered whether the basic RNA unit in TMV might be two RNA chains. While later doodling on paper in my flat, I madly hypothesized that two such chains might be held together by P–O–P phosphotriester linkages. This was a wacky thought because it could only complicate how these RNA molecules were ever copied. But desperate for some way out of our RNA void, I got to the lab early the next morning to see if such a two-chain RNA structure could be built using Pauling–Corey space-filling models.

Straight away, I found myself putting together a seemingly perfect structure where all the atoms fit snugly together with just the right interatomic distances. Happily, there was a hint in the literature of unstable TMV structure becoming more acidic upon breaking down. Although these were crude 1937 data, my new model could be quickly proved or disproved by isolating RNA from TMV in the presence of ^{18}O-containing water. If phosphotriester linkages existed, then ^{18}O molecules should be found in the resulting purified RNA. But I realized that getting the answer might take many months to several years.

Previously I had taken my elegant-looking two-chained ribbon to the Chemistry lab for Linus Pauling's longtime coworker, Robert B. Corey, to look at. He had to admit it was indeed pretty, but would aesthetics win? Corey's "American Gothic" demeanor went along with his long-subservient role to Linus. I was surprised, therefore, when he candidly implied that Linus, then touring India, was far too out of touch with his experimental helpers.

Total frustration with his meson theories was bringing Dick Feynman to my office most afternoons. Recently he had spent an entire evening staring at his ceiling trying to make sense of the data that seemingly had no rhyme or reason. An unexpected invitation to go to Moscow for an "All Union Conference on Quantum Mechanics Electrodynamics" initially diverted him, until he realized that the State Department might not want him on the other side of the Iron Curtain. Close to the date that he was due to depart, his passport was withheld.

My morale increasingly was on a roller coaster, the inherent shakiness of my ideas particularly hitting home when more than a week would pass without a letter from Christa. But I was kept sane for the lectures I was then giving to a Virus course by the prospect of flying East to give a March 17 lecture in Baltimore at the yearly Biophysical Society Meeting. Not at all sure what would happen when I saw Christa afterwards, I penned another letter to Foxcroft School in the hope that Margot Schutt and I might meet in Washington. Its cherry blossoms would be out and a perfect time for the aesthetically tuned Margot to see that I go for more than molecules. These hopes, however, never went anywhere. On the day I was to speak in Baltimore, she would be in New York seeing "mommy and daddy" sail for a spring in Italy. Then she was going on to Boston to help decide whether to go to graduate school or seek a career in publishing.

My trip took on a completely new dimension when, just after my arrival East, I got word that the Harvard biology professors had voted to offer me an Assistant Professorship starting on July 1. Before they made up their minds, they had sent out letters to several prominent scientists asking, among other matters, whether the low voice I used in my job seminar might later be equally inaudible to Harvard students. One such letter went to Fritz Lipmann, long notorious for giving such incomprehensible lectures that the Harvard Chemistry Department had appointed Konrad Bloch, not him, to introduce biochemistry into their undergraduate course offerings. In reply to the botanist John Raper's letter, Fritz wrote "that if I have something important to say, the students will hear me." Soon after receiving the letter from Frank Carpenter, the Biology chairman, I quickly let him know that I would end my East Coast visit at Harvard to see its laboratory space and facilities. Of course, I would accept his offer.

20

The East Coast, Pasadena, and Woods Hole: March–June 1955

WHEN I ARRIVED in Baltimore for the Biophysical Society Meeting I found Geo Gamow keen to talk about the new manuscript we had each separately received from Crick. In the 17 mimeographed pages entitled "On degenerate templates and the adaptor hypothesis: a note for the RNA Tie Club," Francis conveyed a radical new thought about how the genetic code works. Never intended for actual publication, it reflected Francis's opinion that neither DNA nor RNA had the structural features that would let them act as direct templates for the ordered assembly of amino acids into polypeptide chains. Instead he regarded nucleic acids as molecules that like to form hydrogen bonds but not to make the hydrophobic interactions necessary for distinguishing the water-repelling side chains of amino acids such as valine or leucine.

Francis first focused on this dilemma when driving early in September 1954 from the Nucleic Acid Gordon Conference to New York City, to sail back to England and his family. On the road, he clicked onto the bold idea that each amino acid, before its incorporation into a protein, is chemically bound to a small, possibly RNA-like, molecule with a specific hydrogen-bonding surface that in turn binds specifically to sets of RNA bases. When learning of this idea in December, Sydney Brenner called it the "Adaptor Hypothesis" because it proposes that each amino acid is fitted with an adaptor to attach it to the template. In its simplest form, there would be 20 different kinds of adaptor molecules, as well as 20 specific enzymes to join the amino acids and their respective adaptors. Although no such small adaptors had been found, Francis argued that they could easily be present in small amounts and had been overlooked. Furthermore, the energy that must be supplied to join the amino acids together to form peptide bonds might be supplied by bonds uniting the adaptors with their amino acid.

Geo saw no reason yet to accept Francis's reasoning, but he was not averse to considering it—better a still-untested new idea than an old one going nowhere. In contrast, I did not like the idea at all. If correct, it meant that Leslie Orgel and I had been naive chemists when unsuccessfully we tried to fold RNA chains into helical structures with hydrophobic cavities complementary in shape to the hydrophobic amino acid side chains. More to the point, the adaptor mechanism seemed to me too complicated to have ever evolved at the origin of life.

On the other hand, I had to admit that we had never come close to making a helical model where the hydrophobic backsides of the purine and pyrimidine bases formed even the simplest of cavities—say one into which the methyl side group of alanine would snugly fit. So maybe the first real meeting of the RNA Tie Club should be devoted to the merits of the Adaptor Hypothesis. But now the club's National Science Foundation (NSF) funding was problematical. Those in charge wanted to be sure that those invited would actually show up in Boston. Much worse, they wanted a book out of the meeting to justify the $3500 we had asked to cover travel and subsistence costs. Common sense told us that more hassle than pleasure would come from further pursuit of such funding.

What the next day would bring with Christa was never far from my mind. I sensed that all might not be perfect when I learnt that she would be coming down on Saturday morning, not on the Friday afternoon when her classes at Swarthmore were over. I went back to Washington with Geo to spend the night with the Riches. Not knowing what Christa was up to kept me from sleeping well and I was still worrying when she came off the platform carrying her overnight possessions on her back. The kiss I wished to give her evaporated into an awkward hug. As we rode the trolley out to Georgetown, I talked as neutrally as I could about Geo's reaction to Francis's adaptor idea.

My objective in Georgetown was an arty bookstore. After entering I stared at a large Gauguin reproduction of Tahitian women. Before then,

<u>ON DEGENERATE TEMPLATES AND THE ADAPTOR HYPOTHESIS</u>

F.H.C. Crick,

Medical Research Council Unit for the Study of
the Molecular Structure of Biological Systems,

Cavendish Laboratory, Cambridge, England.

A Note for the RNA Tie ·Club.

"Is there anyone so utterly lost as he
that seeks a way where there is no way."

Kai Kā'ūs ibn Iskandar.

In this note I propose to put on to paper some of
the ideas which have been under discussion for the last year or
so, if only to subject them to the silent scrutiny of cold print.
It is convenient to start with some criticisms of Gamow's
paper (Dan.Biol.Medd.22, No.3 (1954)) as they lead naturally
to the further points I wish to make.

Some straightforward criticisms first. The list
of amino acids in Table I of the paper clearly needs reconsider-
ation, and this brings us to the very interesting question as to
which amino acids should be on the list, and which should be
regarded as local exceptions. We first remove norvaline which
we now know has never been found in proteins. Nor, as far as
I know, is there at present any evidence for hydroxy glutamic
and cannine. On the other hand asparagine and glutamine
certainly occur, and indeed are probably quite common. We now
come to the "local exceptions". These are:

$\Big\{$ hydroxyproline
hydroxylysine

$\Big\{$ tryosine derivatives, i.e. diiodotyrosine,
                                          dibromotyrosine
thryoxine, etc.

diaminopimelic
phosphoserine.

The first two occur only in gelatin. The tyrosine derivatives
are found only in the thyroid (the iodo ones) and in certain
corals (and in other marine organisms?). Diaminopimelic
occurs only in certain algae and bacteria and has not yet been
shown unambiguously to occur in an ordinary protein.
Phosphorous occurs in casein, ovalbumin and pepin, and may be
present as phosphoserine.

There are, in addition, amino acids which occur in
small peptides, such as ornithine, diaminobutiric,etc. - see
Table I of Bricas and Fromageot, Ad.Prot.Chem.(1953) Vol.VIII
for a comprehensive list. Under this heading one should also
include the D isomers of common amino acids, and ethanolamine,
which occurs in gramicidin.

-1-

Christa's face had seemed to me a Renoir painting, but from this moment on I saw her as a Tahitian beauty. My saying so, however, did not have my desired effect. At lunch in a nearby coffee shop, neither of us was at ease and soon we were back outside for a desultory walk along Wisconsin Avenue. We arrived sooner in Cabin John than the Riches expected, and passed more awkward moments walking several miles along the towpaths of the Chesapeake Canal. It was all too clear that my honest, twice-weekly letters to Christa had backfired. Better would have been brisk reports of my activities, not outpourings of unhappiness at her absence.

During dinner Alex and Jane sensed that our weekend was not on course, particularly after Christa told them that unexpected homework required her to be back at Swarthmore by Sunday suppertime. My psyche took another hit the next morning when she said that she could easily get back to Union Station by herself. Then Alex came to my rescue by offering to take all of us there in his car, thereby letting our lunch be less hurried. Only on the platform walking her to her coach did I sense a lightening of her mood. Then to my surprise, I realized that our past plans for her to come up to New Haven the following weekend were still on course. We were to stay at my uncle and aunt's house just off the Yale campus. Given my aunt Betty's longstanding preoccupation with dead ancestors, the visit might be the emotionally neutral occasion I badly needed.

The following morning I thought I made a good impression on Lawrence Blinks, the NSF biology chairman, whose good graces could help bring about the NSF grant to let me return to England. Most important to him was that I would soon have a Harvard position. This should allow the NSF to give me a grant as a member of the Harvard faculty that in turn I could use to fund my time in Cambridge, England. I still had to convince Harvard's Biology Department that my academic appointment start with a NSF-funded sabbatical-like year. But a week later I found its chairman, Frank Carpenter, happily agreeable to the first year of my five-year appointment being spent abroad. With NSF funds for my time in England, Harvard would save the money it otherwise would have to use for a later, more conventional sabbatical leave.

Only a little Harvard departmental money would be available to let me equip and run my lab a year hence. But I anticipated no difficulty in getting more NSF support that would let me do science at full tilt. The only substantial Harvard issue to settle was the space in the Biological Labs to suit my position as an Assistant Professor. Because several major professors were soon retiring, there was free space galore, even excluding the unfinished north wing. Several large, square, corner offices were out of bounds except to full professors, but I was shown the almost as large, third-floor office soon to be vacated by Frederick Bailey, long revered as America's best plant anatomist. No chemical hoods were present in his accompanying microscopy lab, but I still saw great advantage in taking over his modest facilities. The office faced west out over large trees and grass to the architecturally pleasing nineteenth-century Divinity Dormitory. Moreover, it was just above the formal entranceway flanked by two life-size bronze rhinoceroses that, in 1934, the soon-to-retire President Lowell had judged appropriate symbols of biological research.

In New Haven with Christa, it was as if the past weekend had never occurred. Why she was less tense, and so naturally put her arm around me on my aunt's newspaper-strewn couch, I had the good sense not to ask. Christa's natural good looks were soon noted by my uncle. In turn, my aunt took comfort in reporting to friends that my girlfriend was the daughter of a prestigious Harvard professor. And I now had the warm inner glow from Christa happily telling me that from Swarthmore she would be mailing me an appropriate present for my forthcoming birthday on April 6.

The thought of her gift kept my morale high through the tiring journey back to Caltech. Needing to change planes in Chicago, I briefly saw the Orgels—Leslie back again being a pure chemist with Robert Mulliken. When my birthday passed without receipt of Christa's present, I grew anxious until I saw her writing on a long cylindrical parcel. Savoring its opening until I was back in my flat, I found the reproduction of Gauguin's *Ta Matate* that I had so admired in the Georgetown bookstore. With its three Tahitian women's faces looking wonderfully serene, I placed it in a prominent place on the wall above the table that I used as my writing surface.

Afterwards in carefree spirits I went off with the Delbrücks, the Dulbeccos, and their German-born friend, Marguerite Vogt, then working on polio with Renato, for an Easter camping trip. Our objective was the desert sand dunes on the eastern side of Baja California, just south of the Mexican fishing village of San Felipe. The heat was not quite that of the parching summer, but the grunion-filled water was already tepid and Manny Delbrück, still girlishly formed at 36, saw no reason for swimming suits. Modesty initially made me keep my head largely under the water, but Max and Manny showed no such inhibitions nor did Marguerite, whom I found particularly awkward to view at close distance. But by dusk, it was obvious to everybody from my burnt skin that I had not been completely modest. A week passed before my fried look vanished.

Now that I had formally accepted Harvard's position, I could enjoy looking forward to a brief visit from George Wald. It soon was punctuated by an awkward moment when Max, in an uncharacteristically gentle manner, pilloried the mathematics behind George's new theory of how the eye adapts to light. Although little time remained for me at Caltech, I still wanted a serious manuscript to emerge from this year's academic efforts. By then, however, both Leslie and I felt our triplet scheme for RNA synthesis too speculative to publish, and I knew that it would be likewise dangerous to announce my more recent RNA ribbon structure unless future experiments with ^{18}O-water demonstrated RNA phosphotriester bonds. For the moment, I felt semi-comfortable with the theoretical paper about tobacco mosaic virus (TMV) RNA that Don Casper and I had started writing. Using symmetry arguments for the existence of protein subunit dimers, I first favored the existence of some 12 RNA two-chain ribbons running the whole length of the 2800-angstrom-long TMV particles. But it implied an RNA chain molecule weighing only half the 200,000 value measured over the past month by Norman Simmons in Los Angeles. So we might not yet have a good paper in our grasp.

Gunther and Inga Stent had urged me to visit Berkeley one last time before leaving California, and I chose a late-April date that coincided with my sister Betty and her husband Bob's arrival in San Francisco by ocean liner from Japan. Proudly they showed off their almost one-year-old son, Timothy, while I expressed excitement about spending another year with Francis in England before having to start teaching at Harvard. Betty and Bob had been living well as part of the U.S. government's occupying force in Japan, successively in two stylish homes—one built for a member of the powerful Mitsui family near International House in Tokyo, the other to the south in a beach home near the Great Buddha at Kamakura. What Bob had done for whom in his office at our big naval base in Yokohama, I saw Betty didn't want me to ask.

After putting them on a plane for Chicago, I crossed the Bay Bridge to Berkeley to give a seminar. I first talked about RNA ribbons and then, even more speculatively, about possible DNA structures generated during genetic recombination. That evening Gunther reported rumors that RNA had recently been enzymatically synthesized in the New York laboratories of the Spanish-born biochemist Severo Ochoa. Marianne Grunberg-Manago, a postdoc from Paris, had apparently found a bacterial enzyme that made RNA chains using nucleotide diphosphates as precursors. DNA was said to have no role in this reaction, which generates RNA chains with random sequences of the four different nucleotide building blocks. The absence of a role for DNA made Gunther and me wonder whether the true cellular role for this enzyme was the breakdown, not the synthesis, of RNA. In any case, we were keen for Alex Rich to get his hands on this enzymatic product soon to see whether it gave the RNA X-ray diffraction pattern that we had found common to both viral and cellular RNAs.

Back at Caltech, my last weeks passed quickly. Even a brief stench of awful smog failed to dampen my spirits. One weekend, I went into the Mojave Desert with Matt Meselson, who just had been given a Thunderbird convertible by his affluent father. With ease we sped along at 100 mph, and as we drove back I sensed that an open roadster would greatly enhance my forthcoming return to England. For several days, Robert Oppenheimer set the Caltech campus abuzz with his two lectures on meson theory. Without any hesitation, I joined the curious mob that packed the Physics lecture hall and like most was not qualified to follow his arguments. But his lucid mannerisms mesmerized many of us into temporarily believing we knew what he wanted his theory to achieve. The day before his lecture, his porkpie hat silhouette suddenly cut across my path to the Athenaeum, his face so familiar that I felt as if he had been part of my personal world for a long time.

An infinitely less appealing treat was several hours of Erwin Chargaff. His May visit was his first to Caltech, and in no way did he compromise his still-vitriolic feeling towards Francis and me. Referring to us as "The Thinkers," Dick Feynman kept jabbing me in the side as Chargaff rose to new heights of disdain. Newcomers to DNA might have suspected that Chargaff's bilious remarks reflected his failure to be the first to appreciate the significance of his $A = T$ and $G = C$ base ratios. I knew otherwise, remembering well his petulant attitude to Francis and me in John Kendrew's Peterhouse rooms the summer before we found the double helix.

For my last two California weekends, I went with friends to the mountains to the east of Caltech, first going up the 11,500-foot San Gorgonio and a week later reaching the slightly higher San Jacinto peak above Palm Springs. There was still snow on both peaks but not too much to confuse our paths and make us wish that we were back in safer lower elevations. Without much formality, I was effectively closing out my Pasadena existence. Saying good-bye to George Beadle was the most painful moment. In every way possible, Beets had been on my side. But knowing that my long-term self-interest likely lay at Harvard, he was invariably gracious when my departure came up. In the 1930s, a year of Harvard had been one too many for his Wahoo, Nebraska-bred mind. He now hoped I had the sense not to value Harvard for more than it was.

Four days into June I set off for Chicago, where 72 hours later I let out my co-driver, a biology graduate student, at his parents' house in a North Shore suburb. Then, after a relaxing birdwatching week with my parents, I spent a more intellectual week talking genetics at Cold Spring Harbor before driving on to Harvard. For my first night there, I stayed with the Mayrs in their flat on Washington Avenue. Christa had already left for Europe, initially to wander with a girlfriend through France and then give herself time to become semi-fluent in German before starting her university studies. So Ernst gave me the Fribourg address of her uncle, with whom she would be staying by mid-July, and where I might meet her after I attended a Nucleic Acids meeting in France.

By early afternoon, I was on my way for an overnight visit to Woods Hole where Geo Gamow was spending the summer alone in a small house on the Eel Pond. It let him escape from his nightmarish marital problems with Rho and gave him peace of mind to work on a new book on cosmology as well as more RNA Tie Club antics. Martynas Ycas had already been down from Boston to discuss their scheme for tentatively assigning amino acids to the 64 three-base permutations of A, G, C, and U. Earlier in the year, they had used the nucleotide base and the amino acid compositions of tobacco mosaic and turnip yellow viruses to look for a correlation between the proportions of the 20 different amino acids and those of the four RNA bases. Just recently Martynas had obtained from Berkeley the corresponding compositional data for tomato bushy stunt virus. It seemed to back codon assignments they had made earlier.

As soon as I entered Geo's cottage, he offered me a Scotch to go with the large glass he was already imbibing. But knowing that both of us were going after dinner to a party at the Walds', I deferred drinking until I went nearby to Andrew and Eve Szent-Györgyi's house for supper and local gossip. When we moved on to George Wald's house, Geo was already threatening to limerick-dominate an occasion that George assumed would be for his Brooklyn jokes, best told when his wife was in the kitchen. Geo went through more whisky, but ample beer satisfied the other guests, almost all having returned from the previous summer, and more often than not connected to the physiology course George would again be instructing and where Geo would be a guest lecturer. Near midnight, most certainly Frances and even George were not sorry to force us out.

When I left the party I did not want to be alone for long in my alcoholic haze and persuaded a cheerful, well-proportioned Columbia University student to drive with me to Nobska Beach and its white sands, on which we silently parked. She was both eager for the affection that I had not bestowed last summer and much too nice for me instantly to make clear that something that made no sense last summer made even less sense now. Fortunately she knew that I was flying to London only 36 hours later, which gave me the excuse to start up my car before embarrassing apologies were necessary. For a moment I panicked when the wheels of my car spun in the sand, and I feared that a rescue mission would have to be mounted to push my car back on the road. Then the back wheels gripped better and soon we were in front of the house in which she had rented a room for the summer. She was not at all put out by my passion fizzling and wanted a last big hug to say good-bye. I saw no reason to disappoint her in this.

21

Cambridge (England): July 1955

MY BRAIN REVVED up the moment I was back in old Cambridge in June 1955. The non-hysterical pace of a Caltech oblivious to conversational subtleties was now out of my psyche, replaced by the buzz of Francis Crick holding court within his Cavendish domain. After my first lunch back at the Eagle with Francis, he took me to see Nevill Mott, the clever, now solid-state, physicist from Bristol, chosen the year before to replace Sir Lawrence Bragg as the Cavendish Professor. Until recently, and for almost a year, Mott had erroneously believed that Francis's surname was the middle-class sounding Watson-Crick.

Francis wanted the pleasure of showing Mott his American partner, again to share the first-floor office in the Austin wing where the double helix was discovered. With the pleasantries over, Francis explained how our room was to be jam-packed with visiting scientists, the result of the growing attention accorded to the work of the Medical Research Council (MRC) unit. Precious little free space would be available for the RNA models that we intended to focus on over the next several months. Mott, however, knew better than to rise to Francis's bait by asking how much more space we needed. He knew that we were aware that he wished the biologically oriented MRC unit to vacate its Cavendish site. As soon as possible, Mott wanted the Cavendish completely back to physics and hopefully the preeminent role that it had when first led by J. J. Thomson and then by Ernest Rutherford. But he knew that no satisfactory university space was currently free for the unit to move to and so was resigned to seeing biologists within his province for one or two more years.

My arrival had been preceded several days earlier by that of Alex Rich. He came bearing samples of synthetic RNA molecules made in New York by Severo Ochoa and Marianne Grunberg-Manago. Soon after learning of their discovery, Alex persuaded Ochoa to send a sample of his enzymatically synthesized RNA to his National Institutes of Health (NIH) lab. But the X-ray diagrams obtained generated only marginally acceptable pictures. Very likely the synthetic RNA had a structure identical to that which Alex and I had found at Caltech 18 months earlier. But Alex was not 100 percent sure and saw the availability of the much more powerful, rotating anode X-ray equipment at the Cavendish as a perfect excuse for a short visit to Cambridge. By coming, he told his NIH boss he would have a good chance at last of solving RNA's structure.

Several weeks before, there had been a brief RNA fling in the London papers when they published news from Berkeley that Heinz Fraenkel-Conrat and Robley Williams had reconstituted infectious tobacco mosaic virus (TMV) particles by mixing together purified TMV protein and TMV RNA, neither of which they found infectious by themselves. Aided by University of California publicists, their finding was being touted as a great step toward the eventual artificial creation of life. To vet this hysteria, *The Times* called Francis, who took delight in being quoted that he had been anticipating the successful reconstitution of a virus from its protein and nucleic acid components. It was not a step toward the creation of life but, if reported otherwise, would not have generated the press excitement that Berkeley wanted.

For several days after my arrival, I stayed with John and Elizabeth Kendrew in their little Peterhouse-owned house, 12 Tennis Court Road. John was upbeat about myoglobin—the oxygen-carrying muscle protein—and the effective help that earlier he had gotten from the Midwestern American postdoc, Bob Parrish, and now from the even more determined, Los Angeles–originating Howard Dintzsis. The long fallow period marked by failures to crystallize horse myoglobin happily vanished when John turned the year before to whales and seals as myoglobin sources. Now he and Dintzsis were searching for suitable heavy metals to exploit Max Perutz's 1953 breakthrough that isomorphous replacement techniques should reveal the three-dimensional arrangements of the atoms in proteins.

John was now keen for more serious Alpine walking—like we had done in the Gran Paradiso, north of Turin, in the summer of 1952. For a new adventure, we settled on the first week of August immediately after a meeting on "Les Macromolecules des Vivants" that I was to attend at Pallanza (Verbania) on Lake Maggiore, north of Milan. Francis, not me, was initially asked to speak at this non-mainstream gathering, the proposed program for which held out little hope for intellectual enlightenment. At first Francis smelled a practical joke perpetuated by Italian friends of mine and so needed reassurance that the gathering was for real, and might help put Italian biology firmly on the DNA track. But with Alex Rich on the Cambridge, Mass., scene, Francis saw no point in taking off a week for an affair where the food and scenery, not the science, was to be the main attraction. He could back out easier if I took his place on the program, and I readily accepted his offer. Being there, I would have reason for crossing Switzerland into Germany to see Christa Mayr.

From Clare College's head tutor I got permission to occupy, until the fall term started, a garret room in the Old Court adjacent to the Cam. Out of its late-seventeenth-century windows, I could look over the classical features of Clare to the splendors of King's College Chapel. That my bed was granite-hard, that no washbasin was nearby, and my bedmaker was uncommunicative were irrelevant. Only some three minutes were needed to reach Trinity Street, buy *The Times*, and nip into The Whim for breakfast. Later I could stroll up King's Parade, pass the Senate House, and only two minutes later walk through the gated passageway of the Cavendish Laboratory.

Technically I was again a research student coming back to complete a Ph.D. thesis not finished because my draft board dictated my return to the States. Through this dodge I could live in Clare with the small university bench fees paid out of National Science Foundation funds that Harvard was passing on to Max Perutz to cover my research costs at the Cavendish. The remaining $875 was sent to Caltech for a large set of Pauling-Corey space-filling atomic models. When they arrived, Francis and I could have a fresh start in assessing options for RNA's structure.

Until then we planned to complete our unfinished theoretical article on plant-virus structure, contrasting the helical symmetry of TMV with the probable cubic symmetry of spherical viruses such as turnip yellow mosaic virus (TYMV) and tomato bushy stunt virus (TBSV). In it, we would emphasize the structural consequences of the viral protein shells being constructed from smaller protein building blocks. Soon we would be joined by Don Caspar, who had just obtained funds that would let him be for a year at the Cavendish before taking up a research position at Yale. Here in Cambridge his objective was to use the unit's powerful X-ray equipment to establish unambiguously the cubic symmetry of the protein shell of TBSV.

I had hoped to dine with Hugh Huxley often, but his experiments never seemed to generate free evenings for socializing. Now back from the Massachusetts Institute of Technology (MIT) for almost a year, he was very excited by the sliding-filament model of muscle contraction implied by his MIT electron-microscope results. With his Ph.D. finished, he had the total independence that came from his recent appointment to the MRC unit's staff. Upon Hugh's return, John Kendrew initially hoped that Peter Pauling would become so intrigued by Hugh's new discovery that he would change his research direction away from myoglobin and take up muscle research. Peter, however, made no such move and after my return to the unit, John was resigned to Peter's research, if not his emotional ups and downs, continuing to be much of his own life.

Giving Peter an important part of the myoglobin project was never in the cards, for John could not risk being dependent on his progress. He could never be certain that Peter would actually be in Cambridge, and even when seen coming and going from his Peterhouse rooms, Peter's attention was not necessarily directed to science. When I arrived, Peter was on the continent taking his brother's large, open, touring car to Stuttgart for an engine overhaul by the Mercedes factory. When so fixed, its prodigious petrol consumption might decrease to the point where Peter could actually afford impressing selective Girton and Newnham girls with conspicuous rides through the streets of Cambridge.

The broad Pauling grin on Peter's face that marked his re-entrance into my life several days later was only momentary. His girlfriend situation was also not great. With his old girl flame, Mariette Robertson, now in Paris, he had found it too easy to resume their past Pasadena closeness. But liking Mariette a lot was not keeping Peter from wanting to get equally near to several other full-bodied girls attracted by the Pauling magnetism. Having too many girls to choose from, however, was not the cause of his current discontent. The thorn in his heart was fear that he had made a gigantic mistake in letting Nina, the petite, blond au-pair girl of the Perutzes, slip out of his life and irreversibly return to her native Denmark.

The Pauling charm was also not simplifying the life of Peter's younger sister, Linda, who was now using Cambridge as her base for the year in Europe that had started in the fall. At Reed College in Oregon, she had been a close item with Wendell Stanley, Jr., whose father's Nobel Prize and prominence at Berkeley somewhat tempered her parents' concern that their daughter needed to grow up more before getting too emotionally involved. By graduation time, however, Linda foresaw more excitement in a European year on her own than from early marriage to a potential science graduate student. Happily, her parents proved more than agreeable to the necessary financing.

All went well with Linda's new life until the spring and her growing relationship with her fellow young American, Jonathan Mirsky, then attracted to King's College for postgraduate studies in Chinese culture and history. He was the son of the protein biochemist Alfred Mirsky, then a prominent fixture at the Rockefeller Institute in New York. To Linda's surprise, if not shock, Linus and Ava Helen had reacted badly to their daughter's letter informing them of Jonathan's and her plans for a late spring trip to Spain. To Linda, her parents' objection came straight out of the blue, because they had never before criticized young couples for knowing each other well before marriage. Particularly galling was the fact that her parents personally knew Jonathan's parents. The Mirskys had spent several years at Caltech in the late 1930s, to allow Alfred to bring Linus into contact with the lore as well as the facts of prewar protein chemistry.

Linda could only smell anti-Semitism from two prominent California members of the National Council of Christians and Jews. So outraged, she made her growing Cambridge circle aware of her parents' hypocrisy.

But those in the know about Alfred Mirsky were not convinced. More likely, Linus and Ava Helen did not want their daughter involved with the son of a pretentious scientist, whose ego was inflated far beyond his accomplishments. Mirsky, in the years just before the finding of the double helix, had fought an ungracious losing battle against his colleague Oswald Avery's claim that DNA, not protein, carried genetic information.

The trip to Spain became a matter of honor for Linda. It went ahead as scheduled, though from its start the pair worried that they risked running out of money unless their pesetas were watched. As they drove even deeper into Spain, their shoestring living, far from further cementing them together, was instead aggravating differing emotional needs. Full disaster, however, struck only near the trip's end when they were involved in an accident near the French border that seriously damaged their rental car. Lacking the funds to make the needed repairs, they sent cries of help to England, not wishing to test the response they might hear from Linus and Ava Helen. Luckily, Victor Rothschild rescued them by sending money to get their car back on the road.

By the time Linda somewhat sheepishly crept back into Cambridge, she badly needed a jumpstart. To begin with, she both needed a place to sleep and some form of work to fill her days. Most certainly she did not want to be further beholden to her parents, whose price might very well be her return to California by the summer's end. If only her college years had provided genuine job qualifications instead of manicuring her for eventual marriage to a thinking, probably self-centered, professional man.

Several days later at supper with Francis and Odile Crick, I finally came face to face with her. Cheerfully blond and attractively upright, Linda was not at all hesitant to look you in the eye and say what she thought. First temporarily, and probably now for the summer, she was occupying the front basement room in the next-door house just added to The Golden Helix, the Cricks' home in Portugal Place. With its own entrance, it was a perfect room for an au-pair girl to come and go unobtrusively when not needed to help care for the Cricks' young daughters, Jacqueline and Gabrielle. Odile temporarily was without an au pair and

Linda, not unhappily, was assuming this role. But she wondered how to let her mother know that she was now the domestic chattel of the man who helped deny her father the double helix. But with Alex and soon Jane Rich as fellow housemates, Linda knew her immediate days, though boyfriendless, would not be lonely.

Later that week I dined with Victor and Tess Rothschild at their imposing home of monastery origin on the Cam across from St. John's College. They held court at Merton Hall, the name reflecting long-term ownership of the land not by John's but by Merton College, Oxford. When I arrived, a male servant told me they were still preoccupied with personal business, which gave me the chance of a long talk with their daughter, Emma, whose inquisitive smile belied her seven years. As she showed me about the grounds, I found Emma up to adult babble and regretted when her parents, apologizing for making me wait so long, signaled for her to go to her room.

Over the main course, I brought up Rosalind Franklin's apprehension, learned when I was at her lab the week before, that the Agricultural Research Council (ARC) was looking skeptically on her request for a new X-ray diffractometer. Without it, her research on TMV could not proceed at a pace commensurate with the talents of her new collaborators—an able research student called Ken Holmes and a mathematically powerful South African, Aaron Klug, who was eager to move into biology from his dreary Ph.D. thesis on how molten steel solidifies. Although Victor's role as chairman of the ARC was more honorary than active, he had the ear of its chief administrator, Sir William Slater, and had already been apprised of the additional monies Rosalind needed. Apparently her request would fare better if she made alliances with one of Britain's more senior plant virologists, a step that would waste everyone's time. If necessary, Rosalind was up to this bureaucratic nonsense. On the bad side, no action was likely to be taken until the August holidays were over, but Victor gave me the impression he would quietly be on Rosalind's side.

Although the served dinner was formal, our conversation was far from stiff. It later centered on the fact that Tess, Victor's second wife, came from a long line of ornithologists. Victor's peerage, in fact, was

Outside The Golden Helix—the Cricks' home in Portugal Place, Cambridge, July 1955: (from left to right) Sid Bernhard, Linda Pauling, Francis Crick (with Michael Crick talking to him), Jane Rich, Odile Crick, Jacqueline Crick, an unidentified man, and Ann Cullis; the Cricks' elder daughter, Gabrielle, is in the group of three children in the doorway.

bird-connected. He became a peer only because his Uncle Walter never produced a legitimate heir, devoting much of his life to the mammoth collection of birds that he housed in a special building at Tring in Hertfordshire, just under the Chiltern Hills northwest of London. This devotion to stuffed birds in no way implied a lack of interest in birds of the night, two of which he long kept in residence just outside of Tring. This was not the first time I had heard of Tring. It was Victor's uncle's money that had financed Ernst Mayr's ornithological collecting adventures in the Solomon Islands 25 years previously.

During dessert, Victor wanted to know what Linda was next up to, smiling at the unexpected irony that she was the Cricks' au pair. More seriously he despaired of his inability to help Peter and Linda be more than celebrity brats, abusing privileges that gave them more choices than they could handle. Then, with only a trace remaining of the summer's light over the trees of John's, and with Victor's cognac reeling through my head, I jokingly told Tess I was prepared to put off marrying until Emma reached the age of consent.

I wrote to Rosalind to report on my talk with Victor:

> Cavendish Laboratory
> Free School Lane
> Cambridge
> July 22, 1955

Dear Rosalind

I have had a long talk with Victor Rothschild about your ARC grant. His reaction was very sympathetic and he indicated that he would write Slater immediately. Apparently he knew your situation very well and had been talking with Slater about much new additional funds you could usefully absorb. The idea of having one or two people attached to K. Smith's group appealed to him, and I think that Bernal should submit a detailed proposal along these lines. However, with August holidays coming up, nothing is likely to happen until Sept. Perhaps it would be a good idea if we could talk again before anything positive is done, so that the mistake of applying for too little could be avoided. Unfortunately I plan to leave for the continent on holiday on Monday and so it's not likely that I shall see you before Sept 1.

I shall most likely stop in Tubingen on my return from Switzerland

> Jim

As July progressed further I—often with either Peter or Linda—saw more and more of the Swiss biochemist Alfred Tissières, who as a research fellow of King's enjoyed a splendid first-floor Gibbs Building suite almost next to the Chapel. Late one afternoon, Linda, Alfred, and I happily went off to Victor and Tess Rothschild's annual July garden party where ample punch more than compensated for the lack of other younger guests and kept us light-headed through the last vestiges of long-lasting, bluish twilight. A Bentley owner, though now it was temporarily on blocks at his mother's home in Lausanne, Alfred was an experienced Alpinist who had been to the ice falls of the treacherous Rakapushi of northern Pakistan. He had spent the first part of his research fellowship at Caltech when Francis and I were searching for the double helix. Now his research in the Molteno Institute centered on the cytochrome proteins under the umbrella of David Keilen. In August he would be back in Switzerland and, after my walking tour with John Kendrew, I planned to join him in the Valais at Zinal.

Alfred was but one of many distinguished Alpinists with King's connections. One was the once-powerful economist Arthur Cecil Pigou, whose limp, almost 80-year-old emaciated body was draped on most sunny afternoons over a folding beach chair in front of the Gibbs Building. There Pigou noticed that Alfred, this summer, was frequently with a young woman, whose mannerisms augured more than friendship. So Pigou, whose misogyny was extreme even for King's, worriedly took Alfred aside to tell him that a woman's place was in someone else's home. Even more important, they were the enemy of the hills.

Before I went to the continent, a letter from Geo Gamow arrived. It was the first official RNA Tie Club circular, written on the club's new stationery. The 20 club members were listed together with their amino acid code names. Also listed were two potential honorary members, Albert Szent-Györgyi and Fritz Lipmann. Geo proposed that four honorary members eventually be elected but wanted to collect first the monies needed to provide the first two members with club ties and pins. The $2 that he wished to extract from each real member, however, later proved a stumbling block to the honorary category. As club officers, Geo put himself at the top of the list as the "Synthesiser," while I and Francis were the official "Optimist" and "Pessimist," reflecting our differently held enthusiasm for RNA's potential to bind specific amino acids. The Lord Privy Seal of the British Cabinet had no special interest to plead, but why Geo gave Alex that designation was a mystery. It was easier to understand Geo's choice of the avid data classifier, Martynas Ycas, as the "Archivist."

Geo's catchy club motto "Do or die or don't try" hit me squarely in the belly. We could never say, "Don't try RNA." That would mean giving up further pursuit of the gene. Better to die scientifically than not go into this battle. Human affairs were another matter. Although it was too late to say don't try Christa, dying because of her was not the way I wanted to go.

The Continent: August 1955

CAMBRIDGE HAD SETTLED into mid-summer quietude—except for the tourists all bound for King's College Chapel—when I left for my month's stay on the continent and the macromolecule gathering just across the Alps at Pallanza (Verbania) on Lake Maggiore. A brief plane flight across the Channel to Paris let me spend my first night with Boris and Harriet Ephrussi in their new home near the new Gif laboratory to which they would soon move their research groups. Buying their modest home had exhausted their free monies, and I went to sleep on a little day couch extended with a chair to accommodate my 6-foot, 2-inch length. Over supper Boris talked about the offer he had just turned down to move to the University of Chicago. With Harriet expecting their first child, they did not want to endure its decaying urban environment. Harvard, however, would have offered a real change from Paris, especially as Harriet was American and 20 years younger than the previously married, Russian-born Boris. Both had enjoyed their four-month sabbatical in the Cambridge next to Boston, but to their sorrow no hint was later given to them that Boris was wanted back to replace the still academically anemic Paul Levine.

Over supper my throat was beginning to feel raw, and I awoke with it so sore that swallowing was an ordeal. I, however, saw no choice but to press on with my journey. After getting a tour of their new research facilities, I was driven to nearby Orly and the plane that was to take me on to Geneva. There for a day I was to be the guest of Jean Weigle, now back from his Caltech winter in the Delbrück orbit and again ensconced in his Geneva lab doing more phage experiments with his former student, Edward Kellenberger. Of Genevan heritage and of sufficient financial means to let him now have no teaching responsibilities, Jean was a keen Alpinist and listened with interest to my report on Alfred Tissières's current life as a fellow of King's College. They had previously moved in the same French-speaking Swiss climbing circles, but Jean, now aged 55, let me know that the 17-year-younger Alfred could do ascents that he no longer attempted.

The cliffs of the Salève, the mountain over the French border, on Geneva's southeast side, was the training ground where Jean learned how to face the high voids separating him from the ground many hundreds of feet below. His pleasure in climbing came not only from mastery of his toes and fingers on the rocks but also from muting the fear of falling that he admitted was often with him. Two years before, Jean had led me roped up to the top of the Salève, a minor feat that left me exhausted yet optimistic that when roped to experienced climbers, I might get to the top of one or more real Swiss mountains. Jean that afternoon, however, was less keen about climbing than about his newest experiments concerning the bizarre changes in phage lambda's hereditary properties that accompanied its growth on different bacterial hosts. But soon sensing that my throat was so sore that I could barely reply, I was quickly seen by a doctor who told me that I needed a course of penicillin and several days of rest to cure me of raging tonsillitis.

My illness created the dilemma of where I was to recuperate. Jean had expected me to stay only one night in his austerely elegant apartment on the Place de Mezél, down the Grand Rue from the massive Cathedral dominating the old city of Geneva. Then my guestroom was to be occupied by an older woman friend coming from France, who Jean let slip was of Rothschild ilk. I had to go, but at first I did not realize my good fortune when Jean told me that my recovery would be presided over by Ann McMichael, the blond wife of a young American physician from Philadelphia undecided between pure research as opposed to clinical research. The McMichaels' visit to Geneva was a follow-up to one made two years before to Caltech, where Ann had spent much time with Jane Rich while their respective husbands were in their labs.

That I might not want to get well too soon hit me the moment the McMichaels arrived the next afternoon to take me to their small hotel on Lake Geneva. Ann had the good looks and warm personality of a typical American college girl that had so eluded me before on American campuses. Immediately she waxed warmly about their hotel's marvelous vistas across Lake Geneva to the mountains on the French side. Later, eating the delicious pastries that awaited us at the hotel, I began to feel much better, especially when I noticed her gaze showed no tendency to turn when my eyes were simultaneously focused in her direction. By the next morning the penicillin had already done its job, and with my sore throat and fever virtually gone and with her husband already off to the lab, we wandered in and out of the lakeside shops. After our pension lunch at the hotel, we rushed to catch a tourist boat that went over to the French side. Soon, almost as new lovers, we walked and looked at each other among nearby orchards and vineyards, enjoying our luck in my being too sick to go on to Pallanza. Although I was now well enough to reach the meeting before it ended, my telegram had gone off the day before and no one was still expecting me.

After we returned to the hotel for dinner, I started feeling guilty about not now trying to attend a small RNA meeting that had just started in France, only an hour's drive south from Geneva. Six months before, Paul Doty and I helped organize it with Henri Lenormant, who had recently worked at Harvard Medical School. Initially the proposed tiny "RNA and Protein Synthesis" meeting in Lenormant's provincial hometown of Culoz sounded like pure fun. To be sandwiched in between two international congresses, one of chemistry in Zurich, the other of biochemistry in Brussels, it might substitute for an RNA Tie Club meeting on adaptors. But Lenormant had no more success in getting modest French monies than Geo Gamow did with American funding. Then a grant of $150 from the Rockefeller Foundation gave the meeting its final form. Only the food and some accommodations at Culoz could be covered by this tiny sum—all travel costs had to be borne by the participants. Thus when I was later offered Italian funds to bring me to the continent, I decided to go to Pallanza instead of Culoz. It would have been impossible to attend both meetings since they were to be held at the same time.

Ann McMichael now saw I could get to Culoz semi-honorably if, the following morning, she and her husband brought me there as half-sick in their car. The weekend was starting, and we drove off with the hope that they also spend Saturday night at the Chateau de Béon, the large home of Lenormant's friend, the Baron d'Aiguy. There all of the participants were housed, the meeting itself being held at the local school. Only some 12 persons actually had gotten to the meeting, and so there was no trouble in inserting me onto the program to give an afternoon talk on tobacco mosaic virus. Starting with my 1952 Cambridge effort that established its helical symmetry, I later focused on its RNA component and whether there was a single or multiple identical chains enclosed within TMV's coat of helically arranged protein subunits.

Two unoccupied rooms were then found to give the McMichaels and me the opportunity of staying on for a most memorable post-meeting banquet of food and wine with much animated conversation that increasingly was in French. After dessert, with Ann's husband seemingly content talking science, Ann and I followed the Baron down to his wine cellar to obtain more of the Sauternes that we were raving about. There he took pleasure in opening up one of his more cherished bottles that the three of us soon made short measure of. By then, Ann and I found it much easier to sip good wine rapidly than to reveal further our inability to comprehend even the limited repertoire of the French words that our gracious host dared inflict on us. Soon alcoholically unable to think ahead more than the next minute, we slipped out into the garden and the smell of roses that made us feel close and warm. Suddenly, realizing that no more voices were coming from the party, we went back inside and I was soon asleep in the sparsely furnished upstairs servantlike room.

The following morning I was driven back to Geneva to catch the train that would let me meet up with John Kendrew for our week of cross-Alps walking. Ann's affectionate smile deepened only slightly as we said good-bye, and her husband helped put my rucksack inside the train doors. My return to England would take me three weeks later through Geneva, but the McMichaels' plans might be taking them away that weekend. Ann's good-bye wave told me she hoped otherwise.

John Kendrew at Peterhouse

Three hours later, I went through the Simplon Tunnel that might have taken me earlier to Pallanza. I got off one stop earlier at Domodossola to catch the bus up the long valley to Macugnaga, the Italian mountain village lying below the Monte Rosa, the tallest mountain in the Alps. There at the small prearranged hotel was John, beginning to worry over why I had not shown up in time for supper. The next day we walked up to the relatively low Monte Moro Pass (2832 meters), from which we could later walk down to Saas Fee, the Valais village across the Mischabel range from Zermatt. It was a stiff 1500-meter turnback hike up to the pass and both of us were exhausted by the time we reached the border post and showed our passports.

Already we were quite reddened by the sun because the day was cloudless and the massive Monte Rosa glaciers gleamed white far above us. In then descending the upper Saas Valley, we lost the trail near the top and had to scramble, if not slide, down a wild scree slope that at first had no visible end. At last we saw the trail to the bottom right and no longer had to contemplate the awful prospect of climbing back up to the pass. Some 90 minutes further down we saw the outline of the small hotel at Mattmark where I wanted to spend the night. But I could not persuade John, who was determined to push on to Saas Fee, some six miles further on. By then the slope was gentle, and I had no further fears of missteps on the rocks. Nonetheless I felt virtually dead by the time we found a simple hotel for the night and barely had the energy to eat our evening meal.

I was so stiff and sore upon awakening that I saw the need for at least a day of newspaper-reading rest before setting out again on the slopes. But I did not anticipate resting alone. To my surprise, John told me over breakfast that he must immediately return to Cambridge. A phone call late the previous evening somehow made his departure imperative. More guarded than normal, John implied to me some most unanticipated event that he must immediately follow up. I was left in the dark as to whether the news was good or bad. In any case, he was soon on the postal bus to Visp and the train that would get him back to England the next day.

Now alone, I wanted the satisfaction of scaling a 4000-meter peak and found a guide who would take me up from the Britannia Hut to the 4027-meter Allalinhorn. The walk up to the hut took four hours, and after supper of cheese and dried beef I went to sleep among some 30 climbers, most prepared for more difficult peaks. It was still dark when we left the hut to go around the Kleine Allalin and onto the huge glacier that rose to the summit of the Allalinhorn. We followed behind a larger party and while the final steps were almost straight up, there was no real chance of a life-threatening fall. At the top, the view in all directions was so exhilarating, I wanted more of a challenge when I was later at Zinal with Alfred Tissières. As soon as we inched down the deep steps near the summit, our pace across the glacier picked up and I was back in Saas Fee before lunch. My next three days were marking time alone in a somewhat rainy Zermatt, reached in less than three hours of postal boat and train time. There my reddened face had turned to brown tan and I almost ran as I came down one afternoon to Zermatt from the Gornergrat.

I felt more than ready for a serious Alpine experience when I moved on to the spectacularly mountainous Val de Zinal. The postal bus initially traversed a tight multiple hairpin road, bordered by huge voids, to get onto the eastern upper shelf of the valley. There my heartbeat calmed and I began to enjoy the mountain views looming ahead. At Zinal, I hoped to find Jane Rich moving on from a lengthy tour of her Scottish forebears. But, instead, I found a Morayshire-posted letter saying she had become entranced bouncing from castle to country home to cottage and finding her relations living in a hodgepodge of grandeur and poverty. So she was prolonging her stay in the land of kilts and porridge.

Alfred, there before me, was soon joined by a Cambridge friend from Trinity College, who sported a long mustache appropriate to his war years in India. After lunch in a simple restaurant, we made the long walk up to the Cabane du Petit Mountet, perched on the moraine wall bordering the lower reaches of the Zinal glacier. We had just started when we saw an English climbing group coming down the trail, and I recognized one of the party as the physicist Willy Seeds, who had worked at King's College London with Maurice Wilkins on DNA during the time of Maurice's impasse with Rosalind Franklin. I thought Willy might stop and chat, but all he said was "How is Honest Jim?" and uninterruptedly continued his downward descent. Alfred saw the irony of the remark and over tea later at the Petit Mountet briefly explained to his friend why the King's lab might regard me as the Judas once close to their midst.

That night Guido Pontecorvo called from his nearby summer home in St. Luc just up the valley north of Zinal. He would join us the next day when we again planned to go up to Cabane du Mountet (2886 meters). From there, with a friend, Ponte was to climb the relatively uncomplicated Lo Besso, in contrast to the major challenges of the Zinalrothorn that Alfred and Michael wanted to ascend. Over supper, Ponte tried to persuade me to do the simpler Besso ascent, but I feared that he was not strong enough to get me out of potential difficulty. At dawn, Alfred, Michael, and I went back down the rocks onto ice and were soon on the massive Glacier de Mountet that climbs steeply up almost 900 meters to the narrow snow-covered ridge leading to the solid rock pinnacle of the

Zinalrothorn. Serious crevices were absent, and without too much strain I virtually reached the ridge before confessing that the final ascent would be too much for me. Alfred and Michael continued upwards out of my view while I turned back no longer terrified by the prospect of what lay ahead. As I came down, the sky quickly filled with dark clouds that increasingly hid the Zinalrothorn and Ober Gabelhorn from my view. Back at the hut, I suspected that the uncertain weather would cause Alfred and Michael to turn back. So I was not surprised when in less than an hour they also appeared. That they had not conquered the Zinalrothorn made me feel less dejected by my loss of nerve. Later I accepted the fact that the real Alps were never for me.

The break from the good weather was prophesied to be long-lasting, and we went down to the flatness of the Rhone Valley and the main line trains running through it. I went east to change to a carriage destined for Basel where I spent the night. The next morning I was on a train going 50 miles north to Freiburg and the relatives' home where I hoped to find Christa. Easily I found the house of Dr. Frederick Simon at Burgunder-strasse 20 and then nervously awaited seeing Christa opening the front door. She, however, was away on a trip and not expected back for two nights. In broken German, I thankfully accepted the offer of a free bed for the nights ahead. After then unsuccessfully using my limited ability to read scientific German to try to converse with her relatives, I retreated to the Penguin version of *Love in a Cold Climate* by Nancy Mitford. Many of its passages made me laugh uncontrollably, and I was totally relaxed when Christa finally walked in the door and unexpectedly found me. Immediately I sensed her eagerness to be with me soon in London and Cambridge and then to go north into Scotland, where I had been invited to spend a September week with Av Mitchison's parents, Dick and Naomi. The next day Christa and I joined her relatives for a long and relaxed walk through the Black Forest and then I was off by train back to Switzerland.

With Christa again on my immediate horizon, I was first disap-pointed and then mildly relieved when, in Geneva, Jean Weigle gave me his "bad" news that the McMichaels were not on hand. Seeing Ann before I knew whether Christa and I would truly pair in Scotland might make me say what I would later regret.

23

Cambridge (England) and Scotland: September 1955

THE LONG VAC was almost over and the tourists were beginning to thin out when I arrived back in Cambridge. The deep-green college quads no longer reverberated from students shouting across to each other, and the coming month would be perfect for contemplation without interruptions. Francis Crick, however, was not in deep thought but happily agitated when I bounded up the Cavendish stairs into our office. Uncharacteristically in before me, he was fiddling with a polypeptide-chain model that I had never seen before. Animatedly Francis explained that it was a model for polyglycine II that an X-ray diffraction article in the latest issue of *Nature* had propelled him and Alex Rich to concoct over the weekend.

They had come up with a stereochemically elegant proposal in which the parallel polyglycine chains, with threefold helical twists, are held together by intrachain hydrogen bonds between the amino and carbonyl oxygens of their peptide bonds. By itself, their solution of the polyglycine II structure was no big thing. It was its implication for the much more important collagen structure that now made Francis and Alex jump with joy. Linus Pauling, five years before, had not gotten collagen right nor had Francis while bored a year ago in Brooklyn. But through their new polyglycine II opening, they might unambiguously nail down the correct structure before the London King's College collagen team put on thinking caps.

Over lunch in the Eagle, Francis shared his recent unexpected letter from my fellow American, Henry Wallace. He had been Vice President during Franklin Roosevelt's third term after serving in his cabinet as Secretary of Agriculture. By now he was generally perceived as a communist sympathizer as he had run for President against Harry S Truman and Thomas Dewey, in 1948, under a Progressive Party banner. But before Wallace got into politics he had been a successful breeder of hybrid corn, and it was as a plant geneticist that he had written Francis about potential applications of the double helix for the plant-breeding world. Unfortunately, we then saw none.

That day I also learned that John and Elizabeth Kendrew's marriage was on the rocks and most likely irretrievably over. The situation was a ghastly mess with Hugh Huxley somehow involved. With that news, I instantly understood John's uncharacteristic bolting in the midst of our mountain walk in August. The matter was just coming to a head when John left Cambridge to join me in Italy. Apparently, upon reaching Saas Fee he learned that Elizabeth had already cleared out her possessions from their house in Tennis Court Road.

That Hugh was in Elizabeth's picture was not a total shock to Odile Crick and me. Three years before there was a big costume party at Roughton House on Adams Road that neither Francis nor John wanted to attend. But their wives thought it would be "the party of the year," so Odile and Elizabeth dressed up for Hugh and me. Odile and I stayed nearly to its end and coming back across Garret Hostel Bridge came upon Hugh and Elizabeth in embrace looking down the Cam. Too much alcohol might have caused this indiscretion, and in any case Hugh was soon off for his two-year Commonwealth Fellowship at Massachusetts Institute of Technology.

The marriage was Elizabeth's second, with her first husband killed when the Cruiser *Dorset* was sunk early in World War II by the Japanese. John was the best man at their marriage before he too went off to war serving in the Middle and Far East as an aide to Lord Mountbatten. In Cairo, he told me he had fallen deeply in love with a Jewish girl whose family stopped their romance because he was not of their faith. When the war ended, he looked up Elizabeth and they found themselves falling in love and in 1948 married. Initially they lived in

Blackheath, just south of Greenwich, to let Elizabeth become medically qualified in London. Even before their 1950 move to Cambridge, John had fallen out of love and did not want to have children that would trap him irreversibly with someone he no longer loved. But he felt responsible for Elizabeth, and also had to worry that he might lose his Fellowship because the then Master of his college, Peterhouse, strongly disapproved of divorce. Until this week, I had no inkling of their troubled life together. But Hugh Huxley's knowledge of their unhappiness dated back to a visit four years earlier to John's art-historian mother, who had long lived in Florence. After the Roughton party, he became Elizabeth's confidant, having already fallen in love with her petite English beauty and intelligence. Elizabeth, though responsive to Hugh's feelings, was still strongly bound to John, and Hugh, unwilling to precipitate a break-up, was glad to remove himself for his two-year Fellowship at MIT.

Elizabeth still remained partly in love with John, though angry at their situation, and hoped against hope that their marriage would reignite. At one stage, she had threatened to leave John and stayed only when he promised things would be different. But their uninteracting lives remained the same. Independently, Hugh, having returned from MIT, and seeing the unhappy situation unchanged, was also in emotional agony, unwilling to intervene yet unable to put aside his love for Elizabeth. He now wanted out of Cambridge again, and hoped then to feel free to get on with his own life, whatever it involved. Unaware of this, Elizabeth had finally decided to make the break. Rumors were now about that Hugh was moving to University College London.

With John's mind not easily focused on science, Peter Pauling saw no harm in soon absenting himself to the continent to bring back to Cambridge his brother's Mercedes roadster. Before going off, Peter had driven Don Caspar to the hour-away Agricultural Experimentation Station in Rothamsted. They had made this trip in a second of his brother's cars, a speedy Porsche that also with time would be shipped to Honolulu. Like the Mercedes, it was in Peter's care. At Rothamsted, Don would get crystals of tomato bushy stunt virus from its plant virologist, Frank Bawden. In 1938, J. D. Bernal had first used X-rays in Cambridge to look at this virus, realizing that its very large size made it then a virtu-ally impossible objective for the X-ray crystallographer. Now Don thought otherwise, and towards this end had just come to Cambridge, arriving on the day that grouse shooting started in Scotland.

By then, I was jittery, anticipating Christa Mayr's impending appear-ance by train and Channel crossing from Germany. The morning she was to come I went down to London and made my way to Victoria Sta-tion long in advance of her arrival. After hugging her on the platform, I led her around Piccadilly Circus and down to Trafalgar Square before taking the tube to Liverpool Street Station for the train up to Cam-bridge. The bowler hats of the "City" commuters together with the coal-gas smells from the fires of the early fall made me feel as if we were almost part of a pre-war movie.

With women officially forbidden to be in Clare overnight, Christa spent her Cambridge nights at The Green Door, the attic apartment in Thompson Lane where the Cricks first lived in Cambridge. It was cur-rently let by Ann Cullis, Max Perutz's efficient, good-looking technician, whose cheerful personality made her fun to talk with at coffee and tea times. Quickly she put Christa at ease by offering us supper after we saw the highlights of the colleges along the banks of the Cam. After peeking in my austere Clare attic room, Christa and I slowly wandered through its Fellows' Garden before going back across the Cam to Trinity and looking up to the overwhelming presence of Wren's great late-seventeenth-century library.

The following morning, Christa got her first view of the Cavendish Lab and Francis at work, and we stayed until morning coffee, hoping in vain that Alex would arrive earlier than usual. After a pub lunch, we acquired a punt and pulled ourselves under the Clare Bridge and up past Magdalene College, happy that the river traffic jams and gawking midsummer tourists were gone until the next year. Over supper at The Golden Helix, Francis pressed Christa as to whether she had come close to falling out of the punt, trapped by holding on to our punt pole stuck in the mud. The next afternoon I rented a bike for Christa so that we could pedal to the north, first following the course of the Cam where the May race "bumps" occur. After passing a lock that lets the Cam drop six feet, we were soon at the Bridge Hotel at Clayhythe for tea. Afterwards, we cycled on to Waterbeach, where a fighter bomber from the nearby American airbase flew low above us, and the temporarily darkened sky and attendant chill over the flat fen farm scene made us relieved that we had brought along sweaters.

Up early the next morning for going into London for a sightseeing flurry, we met Leslie and Alice Orgel for quick, pre-theatre supper. They were just back in England after the Chicago months of their American stay. With Leslie now a Cambridge lecturer in theoretical chemistry, they had moved into a commodious flat carved out of the attic of a large stucco-faced house off Trumpington Road, just beyond Scroop Terrace. We all went to see Claire Bloom and John Gielgud perform in *King Lear* in a Shaftesbury Avenue theatre.

The next day I rented from Marshall's Garage on Hills Road a tiny Morris Minor that somehow was to get us to and from Scotland for our stay with the Mitchisons. Initially the left-hand driving made me quite nervous, but after 30 minutes on the road my only worry was coming out of roundabouts heading in the right direction. We were free to take any road that took our fancy, so the next morning we drove due north 20 miles to Ely and its great medieval cathedral that was just beyond the limits of our past day's biking ambitions. More fen farm views dominated our drive to Wisbech, and soon we were in Lincolnshire, going along flat roads close to the North Sea. The late afternoon had us heading westwards to get around the long estuary of the Humber and up possibly to York. With my legs cramped by too many hours in our tiny car, we quit driving at dusk and, after failing to get a room in a halfway decent town center hotel, found ourselves signing the register at a small hotel on the outskirts of Scunthorpe. Its smoky bar was already filled with boisterous, Saturday-night revelers, whose drunken cries led us to eat quickly in the adjoining dining room. Without testing the outside air, we went soon to our plainly furnished double-bedded room.

With being together no longer a dream but an immediate reality and glazed tired from the day of left-side driving, I became nervous about the moments ahead. After several long fumbles that almost ended in disaster, we soon fell soundly asleep. The awkwardness of our first night in bed never surfaced during the next-day's drive that first took us through York and across the bare moors of Yorkshire and Northumberland. Two hours' passage through the lowlands of Scotland brought us to a small Edinburgh hotel in time for its Sunday night high tea that had aspects of an American bacon-and-eggs breakfast. Then to my acute unease, my first night of fumbles was followed by a second night that was not much better. Only on our third night did our bodies finally move together. We were at the tiny Cluanie Inn on the road that would pass Eilean Down Castle on the way to the Kyle of Lochalsh and the ferry to Skye.

Reaching Skye had long been an ambition of mine since my mother's father was of the MacKinnon clan, most of whose members originated from the southern part of the island. Three years before, when Mother came to see me, we had set off from Fort William on the lower road to the Isles but stopped far short at Glenfinnan when torrential rains did not stop. Now the weather was also rainy but not enough to keep Christa and me from looking up at the jagged peaks of the Cuillin Hills or comparing the flat hills known as Macleod's Tables to the mesas of Utah and Arizona. I kept alert for the name MacKinnon on road signs and increasingly spotted it when we drove through Sleat in Skye's gentler southeast. There we came across a large upended truck against which our Morris Minor seemed toylike.

After a night at a Portree B&B, we retraced our path to Fort William and drove south to Oban, where by happenstance we saw the openings of the Highland Games. We continued down the Argyll coast to the tiny fishing village of Carradale, on the eastern shore of the Mull of Kintyre, across from the lovely mountainous island of Arran in the Firth of Clyde. Here was the Scottish home of Av Mitchison's parents, Dick and Naomi (Nou). As a London barrister, Dick had quarters in the Middle Temple and as a Labour MP needed largely to be in London. But for the Edinburgh-born Nou, then almost 60, Carradale House and the High-land Panel at which she represented Argyll had been the main foci of her life for almost 20 years. In Scotland, she was close to the soil, her cattle, and the problems of her half-owned fishing vessel as well as the occasional poacher who stole salmon from the river that ran through their estate.

The Mitchison land, with its house, farm buildings, and pastures looking down to a sheltered small bay to the south, had been acquired in the mid-1930s to give Nou a Scottish home to complement her Haldane

On Skye with Christa in early September 1955; our rented Morris Minor
dwarfed by an upended truck

blood that had long been a prominent feature of Scotland's intellectual
and commercial life. Born in 1897, Nou had married Dick in the middle
of World War I, which he luckily survived, winning the Croix de Guerre.
During the 1920s their children—Denny, Murdoch, Lois, Av, and Val—
were born, in that order, and raised in a house along the River Thames at
Chiswick when not up at Carradale for holidays and the summer. Carradale House, built in turreted Gothic style more than a hundred years
before, had its large drawing room and dining room on the south sea-
facing side separated by a hall whose large front door went down to grass
that bordered a walled garden to the west.

Besides the main family bedrooms, Carradale House had sleeping
space for various grandchildren as well as guests who came up from
England for the major holidays and summer stays. By late September
when we visited, almost all the visitors were gone except for Murdoch
and his wife, Rowy, still present because their two children, Neil and
Sally, were too young for formal school. I half-expected Av to be there,
but he was back in Edinburgh, where both he and Murdoch did their
research in its university's Zoology Department. But on hand was Av's
close Oxford friend, the Magdalen College mathematician Victor
Guggenheim, escaping from the emotional complexities of less-talented

Carradale House in Kintyre, home of Naomi and Dick Mitchison

colleagues. On our walks up surrounding hills, he and Christa hit it off well, leaving me largely next to Murdoch who was never at a loss for a new observation to explain. Back in the house and over dinner, the talk was dominated by politics and how the Tories were nullifying the actions of the postwar Labour government.

At Carradale House, Nou had the time to ferociously type out one book manuscript after another. Often she did so to purge from her mind emotional demons that, unchecked, might destabilize the narrow line demarcating her role as lady of the manor from the tractor-driving woman of the soil. Already in the mid-1920s, she and Dick had abandoned the emotional restraints that most persons born to their privileges did not know how to live without. Their behavior, in part, was a reaction to the trench slaughters along the Somme that took away so many dear friends of their idyllic pre–World War I years when her scientist brother, Jack (J. B. S. Haldane), and Dick were up at Oxford. Then the still-adolescent and succulently shaped Nou wrote plays or composed pantomimes for them to act in, together with their close friends Julian and Aldous Huxley. Nou proposed to the latter, without success, to have as her lover. Almost blind in one eye, Aldous never saw the trenches across the Channel and after the war remained close to Nou and Jack, whose provocative 1924 *Daedalus* essay about where biology was going became the essence of his schoolmate's 1928 *Point Counter Point.* By then Nou herself was also widely acclaimed as a writer with her 1923 *The Conquered,* which was set in historical times and let her examine the complex bonds of friendship and comradeship. Later Dick and Nou were more into social than political causes, helping to bring the Birth Control Research Committee into existence in 1927. To Nou, who called it domestic prostitution, the conventional notion of marriage (then and now) was profoundly wrong.

Christa was not put off by Nou's strong opinions and occasional outbursts about the inconsiderate behavior of those who disagreed with the way she wanted her house run or the food cooked. Seeing them chatting

*Naomi ("Nou") Mitchison at her son Av's wedding
to Lorna Martin on Skye, August 1957*

together, I was curious about what they were talking about and later was relieved when, unprodded, Nou reported Christa charming though not that formed—witness her being so intrigued by Guggenheim's neurotic meanderings. Did I really want a young bird with wings too fragile for steady flight? These words were not to my liking, and we moved on to Pauling-family gossip and what Linda was like. If she was so lively and good-looking, Nou couldn't understand why Linda and I weren't perfect for each other's needs. Nou didn't expect a straightforward reply, understanding that I felt a need to keep my distance from antics made possible by being part of the Pauling royal family.

The following morning, Christa and I started back to Cambridge, our Carradale stay lasting only three nights because of her need to fly off to Düsseldorf and a further batch of relatives before her Munich University year started. To save time, we stuck to the nasty Great North Road (the A1) and the lorry fumes that hung over our tiny car all the way down to the Huntingdon turnoff for Cambridge. After letting Christa off to spend the night again at The Green Door, I made my dead-tired way to my room at Clare. Going early the next morning to the rental car garage, we were soon on the train to London and the terminus where Christa was to board the bus out to the airport. Undeniably, Christa had thoroughly enjoyed her first introduction to British intellectual life. Whether she had become more or less in love with me, I preferred not to contemplate.

24

Cambridge (England): October 1955

BACK IN CAMBRIDGE, alas, the cold but romantic old courtyard garret room that looked across to King's College Chapel was no longer
mine. The past spring it had been assigned for the coming year to a
third-year undergraduate. I was now ensconced in a modern room
whose windows looked out on the ghastly gray brick of the 1930s university library. When winter arrived, I suspected I would more appreciate it.
Clare's just-completed New Court, built as a memorial to Clare men
lost in the war against Hitler, had centrally heated rooms.

Linda Pauling now also had to move because Francis and Odile
Crick's au-pair girl for the coming year had arrived. Luckily Leslie and
Alice Orgel still had not found someone to look after Vivian, their six-
month-old daughter, born while they were in Chicago. So Linda, unsure
how long her parents would continue funding her to be in Europe, took
up residence in the Orgels' flat at the end of Chaucer Road. A few days
later she was joined by Mariette Robertson, increasingly nervous at the
prospect of spending another Peter Pauling–less year with her parents in
Paris. Now with the Mercedes touring car back in Peter's hands, one to
several Girton girls were likely to get to know him better than she liked
to contemplate. But having made the plunge to Cambridge, Mariette
knew even less where she stood in Peter's eyes. He had just told her that
this fall's Michaelmas term was a do-or-die period for him to get some
real research done.

Two nights later, I brought Av Mitchison to the Orgels' for a Linda-prepared supper. At the last moment, Av had come down from his Edinburgh lab. Although not invited to dinner, I knew Av would be more than welcome—he and Leslie knew each other well from their Oxford days as prize fellows of Magdalen College. As soon as Av and Linda spotted each other, they focused calculated silliness on each other with Av pressing Linda at length about the nature of her au-pair services to the Cricks. Knowing of Av's past crushes for American blondes, one of which had been much too temporary for my sister's ego, I was not surprised when Av asked Linda to join him in Edinburgh. There she could be the au pair for a large flat in the old city that he had recently occupied with his fellow zoologist and vole expert, John Godfrey. To my surprise, Linda instantly rose to the bait. Av's proposal, though initially made in jest, over coffee became a firm offer of a free room, lots of spare time for exploring Scotland or attending lectures, and a weekly stipend of several pounds for covering daily expenses.

To let Linda know what she might face from his Haldane blood, Av asked her to accompany him the next day to Oxford. His formidable grandmother lived in a big 1906 house close to the Dragon School, where his mother, Nou, was educated up to the age of 14. Suitably impressed, Linda all but decided to assume her new duties. She liked the possibility of being within range of a pass by a good-looking scion of one of Britain's distinguished upper-middle-class families. But she was not sure whether Av might be more fearing than welcoming the prospect of her intrusion into his bachelor world. Could his occasional slight stutter be indicative of a brain not knowing how to make a decision?

Waiting in vain almost a week for a direct reaffirmation from Av, Linda finally got the signal she wanted indirectly. In a letter to Leslie and Alice, Av asked whether they would be upset by losing their new au pair. Alice and Leslie, more curious of what Av might do with Linda than worried about their domestic needs, immediately signed off, urging Linda to set off to Edinburgh before Av changed his mind. Departing quickly was also Linda's inclination. Far better to start her new life before her mother got wind of what she was up to and tried to stop her. On the sleeper from King's Cross Station, Linda was nonetheless worried whether the best way to an Englishman's heart was through her cooking.

Four members of the RNA Tie Club in 1955: (from left to right) Francis Crick, Alex Rich, Leslie Orgel, and JDW

In those days, Francis had little time or inclination to talk to me about plant-virus structures. He and Alex were still manic about collagen, wanting to come up with a better cable-like molecule based on three polypeptide chains than the one Linus had proposed five years before. The detailed ways in which hydrogen bonds held the three chains together were what now mattered. Not that they had to hurry because they were in a race with the great Pauling. Once he pontificated a chemical intuition, Linus almost invariably stuck with it. Only with DNA, where his three-chain model was so clearly wrong, had he ever acknowledged a major mistake. Now their genuine competition was from the youthfully pert Pauline Harrison. Linus, in fact, had spotted her several years before as not only pretty but bright when she was Dorothy Hodgkin's research student at Oxford, studying plant-virus crystals. Now in the King's College London lab, where Rosalind Franklin earlier worked on DNA, she was likely also trying to fit polyglycine-like extended polypeptide chains into a collagen cable.

Odile Crick and Jane Rich, finding themselves collagen widows, saw no point in being excluded from their husbands' conversations at The

Golden Helix. Feeling that her time would be better spent at her parents' home, Odile took her young daughters, Gabrielle and Jacqueline, to King's Lynn, some 40 miles north of Cambridge. And Jane flew to Paris to see a New York friend temporarily living there. Until then, I could share with Ruth my Christa anxieties, which rose exponentially as two weeks passed in the absence of a letter from Germany. Nor did I now have any opportunity to share many thoughts with Hugh Huxley. He effectively remained aloof from his Cavendish acquaintances, staying for the most part at Christ's College, where he recently had been made a Fellow. Upon my return from Scotland, I went over to his rooms, but he didn't open up as to where his relation with Elizabeth Kendrew stood. Instead we talked about the powerful new German electron microscope to be put at his use when he moved to London to be part of Bernard Katz's Physiology Department at University College. From others I gathered that the separation of John and Elizabeth Kendrew was final, with no possibility of reconciliation.

I was now spending my afternoons in Michael Stoker's animal virology lab in the Pathology Department. His RNA viruses were likely to offer better systems than RNA plant viruses for understanding how RNA is replicated. In the mornings and many evenings, I was in the Cavendish taking X-ray photographs of RNA-containing potato virus X. It was my response to an overnight visit from Rosalind Franklin, who stayed with the Cricks. Listening to her treat Don and me as insignificant players in tobacco mosaic virus (TMV) research, I felt the need for another plant virus to call my own. Three years before I had obtained a not very good X-ray pattern from a preparation of potato virus given to me by Roy Markham. Now, I wanted photos good enough to establish the helical symmetry of the virus and, from it, the molecular weight of its protein subunit.

The arrival, finally, of a letter from Christa let me enjoy the first real party of the fall, held on a Saturday night in Alfred Tissières's rooms in King's. My head was hazy when I had to climb afterwards into college—I had not been successful in getting an official key to open the door next to the bicycle shed. Even when half drunk, I could hoist myself onto it by standing on a bike below. That evening, as on many others, I dined at Lucy's, a tiny storefront restaurant in All Saints' Passage near St. John's.

At best providing room for only 10 persons cramped at small tables, I had learned of its existence through a friend of my sister's during double-helix days, Geoffrey Bawa. Born in Ceylon to a tea planter, Geoffrey then had been in Cambridge to become an architect after earlier earning a London pre-war law degree. Finding that Lucy's lamb chops and chips still tasted super, I ordered them on virtually all the subsequent 100 occasions I dined there over the next eight months.

Variety in my evening food came largely from the occasional High Table meal in a college. Still an official research student, I did not have High Table rights at Clare and most appreciated a third Michaelmas-week invitation from the Master to join him there as his guest. An experimental physicist, once connected to the Cavendish, Sir Henry Thirkle many years before began devoting most of his time to college duties. His effective long tenure as Head Tutor at Clare meshed well with his bachelor existence. Now portly, but not embarrassingly so, Sir Henry told of the college's good fortune in having Paul Mellon come there from Yale in the late 1920s. Not only had this led to a continuing exchange of students between Clare and Yale, but a major gift from Mellon had made possible the recent completion of the New Clare Court in which I lived.

Sheila Griffiths, my Welsh almost-girlfriend of two and a half years ago, also helped me pass the occasional dinner. After her marriage to Roy Pryce, the young historian she had met in Rome, they had come to Cambridge where Roy edited *The Cambridge Review*. But the month before he had moved to a better position in Oxford. Until she could find an acceptable position in Oxford teaching in a nursery school, Sheila was resigned to living alone during the weekdays in their basement flat off Lensfield Road. With one of her brothers writing for *The Guardian* and her father a Labour Member of Parliament, Sheila's well-intentioned gossip kept our dining hours together free of Christa uncertainties. Afterwards, I could go out to the street without fear of being unable to sleep.

Max Perutz was often dispirited by a mysterious illness that made him weak. It had so long denied diagnosis that most Cavendish denizens had thought it psychosomatic. After a string of doctors had been unable to help him, Max was said to have almost taken up Tony Broad's offer last year to put one of Wilhelm Reich's "Orgone Boxes" at his disposal.

The talented but eccentric builder of the Cavendish's powerful rotating X-ray anode, Tony took Reich's post-Freudian device seriously, saying it might sort out trouble spots within Max's body. Fortunately Max felt less panicky after being told that he might be allergic to the gluten in wheat.

Of light relief to all except Francis, who firmly pronounced himself indifferent, was the Queen's visit to Cambridge in October. Knowing of the route the Queen would take, Jane Rich, who was now back from Paris, and I saw her Daimler glide by on King's Parade. We celebrated seeing the Queen's blue suit by having lunch at the Bath Hotel on Bene't Street next to the Eagle. Often dining there, particularly when he needed to strengthen the faith of a wavering aristocratic son, was the famed Monsignor Gilbey. Easily spotted by his elegant clerical hat, and coming from the wealthy gin family, he was the Catholic chaplain in Cambridge, with an urbane knowing smile taken for profundity by those of his faith. Inherently of more concern to Jane and me, though, was what Peter Pauling was up to. Only on one occasion that fall had I been treated to a ride in his Mercedes roadster. But I later gathered from him that his limited monies for petrol were best utilized getting out to see the girls at Girton, some two miles to the northwest along the Huntingdon Road.

My first glimpse of Girton's inner portals came not through Peter but by way of Linda. Before she left for Scotland in early October, Linda arranged for me to have Friday afternoon tea there with two undergraduates, telling me they might make me less dependent on Christa's whims. Initially they seemed to be more interested in themselves than me, which put me off. Statuesque Janet Stewart was clearly conscious of her good looks, poise, and intelligence. The more down-to-earth Julia Lewis had the sharp features of many English beauties and possibly the need for excitement beyond books and ideas. Soon I suspected that Peter had been expending more petrol upon Julia than she wanted. This latest of Peter's dalliances was probably the reason Linda had me now at Girton. I was about to leave, but then in glided Pamela, the third of the close-knit girl trio. Accompanying her was her visibly possessive friend, Charles Clunies-Ross, whose family still owned the Cocos and Keeling Islands in the Indian Ocean.

Biking back from Girton through the cold darkness of late fall, I went straight to the Cavendish Lab to find Don Caspar. For several weeks he had been frustrated by uninformative X-ray photographs from his bushy stunt virus (BSV) crystals. The day before he finally hit pay dirt by tilting a large crystal at the appropriate angle to the powerful X-ray beam needed to examine objects as large on the molecule scale as BSV. To Don's joy, the resulting photograph displayed the perfect fivefold symmetry expected from a polygonally constructed virus.

Two days later, I accompanied Jane Rich into London. Alex was initially supposed to come, but at the last moment concluded that further conversation with Francis would be more interesting than having tea with Jane's rich aunt at her hotel off St. James's Street. Through Jane, I had been forewarned to expect a self-conscious New York society woman in her middle fifties. I was intrigued to learn that she lived just above the Cold Spring Harbor Lab and had two unmarried daughters who had gone to Smith, the New England women's college that my New Haven cousins had also attended. Conversing with Mrs. Ames, however, proved a far less heavy affair than Jane had predicted. She understood perfectly why my long-term ambition was not to spend all my career at Harvard but eventually to move to Cold Spring Harbor and live in its beautiful, white-painted Director's House.

The next afternoon Jeffries Wyman was about, on his way to Egypt from Harvard, from which he now had irreversibly resigned. To the Dotys' delight, he was leaving most of his valuable antiques with them so that they could appropriately furnish their new, large, mansard-roofed acquisition in the heart of Harvard. After coming to the Cavendish to learn about Max and John's research, Jeffries had tea with John and me at the Regent House. Earlier Max begged off, worried that he might be exposed to gluten-containing food that would magnify his physical weakness. Over scones, Jeffries said I must come in the spring to Cairo where he was to head up UNESCO's new Middle East Science Office.

A few days later, my work with potato virus X took a possibly big step forward, when I developed an X-ray film displaying an almost well-oriented diffraction pattern. If only a little sharper, I might have a real personal trophy to take with me the following week to the Max Planck

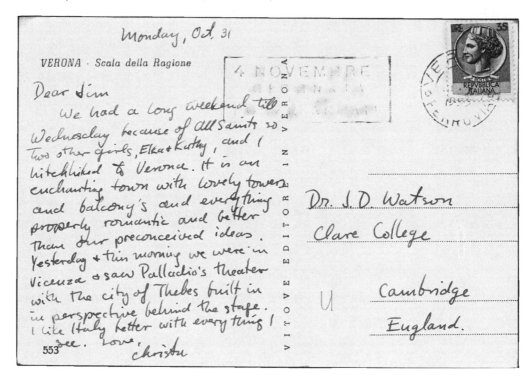

Monday, Oct. 31

VERONA · Scala della Ragione

Dear Jim
    We had a long weekend till
Wednesday because of all Saints so
two other girls, Elka & Kathy, and I
hitchhiked to Verona. It is an
enchanting town with lovely towers
and balcony's and everything
properly romantic and better
than our preconceived ideas.
Yesterday + this morning we were in
Vicenza & saw Palladio's theater
with the city of Thebes built in
in perspective behind the stage.
I like Italy better with everything I
see. Love,
        Christa
553

Dr. J. D. Watson
Clare College

Cambridge
England.

Institute for Virus Research in Tübingen—where the best TMV
research in Europe was being carried out. Even more to the point, I was
going on afterwards to Munich and Christa for a few days. Although her
Munich letters to me ended "With love, Christa," uncertainty about who
might then be near her was always with me.

25

Tübingen, Munich, and Cambridge (England): November–December 1955

MY VISIT TO Munich to see Christa was less than perfect. On my way there, however, my weeklong side visit to the Tübingen TMV laboratory of Gerhard Schramm went more than well. I met his clever young co-worker, Alfred Gierer, who was now isolating TMV RNA. Heading the Virus Institute, though not a bona-fide virologist, was Friedrich Freska, who speculated before the war on gene duplication in an article that had intrigued me while still a graduate student. Before meeting Freska, I anticipated a Max Delbrück–like, tall, precise individual, exuding more questioning logic than warmth. Instead he displayed no trace of arrogance and was delighted with the double helix. I saw him a week later in Munich, when Christa brought me to a public lecture on genetics and DNA that he delivered to a large university audience.

I found myself that week in a no-win situation as I tried to fit into Christa's student life, not helped by the cold raining darkness that enveloped us at most times. I was put off by the bar and student-dominated Schwabing district in which she lived and where I stayed in a worn-out pension for impoverished visitors. Although cranes were up everywhere helping to rebuild Munich at a prodigious pace, much of the center felt as if it had been bomb-shocked. Satisfied living like her largely penniless German equivalents, Christa was more than content with the student food, while I could not hide my need for water to help swallow it down. Mistakenly, then, I tried to bring her to expensive restaurants that she felt she was not appropriately dressed for. All this would not have mattered if I had felt more at ease when we were truly alone. Then my jitters made me less than I wanted to be. Not helping these moments were Christa's fears of the past weeks that her heart was beating irregularly. To her displeasure, I wanted her to see a specialist to find out whether there was anything more than the "nervous Mayr heart" that had bothered her father in his stressful early university days.

As we walked through the Alte Pinakothek, Munich's grand art museum, I kept hoping that Christa would share her feelings about its paintings, particularly those from Italy with serene Madonna faces. But she gravitated towards the big Kandinsky oils whose asymmetrical curves I found jarring and not conducive to bringing couples together. At an evening concert featuring Bach and Mozart as well as Carl Orff, we smiled but all too soon went back out into the cold evening damp and my stomach was tight as we got ourselves back to Schwabing. On my last Munich morning we went into the city center to find a tourist-oriented shop from which I could post Christmas presents back to my parents and sister. Christa here took charge and picked out some Bavarian wicker baskets. Not much more than a quick kiss of her cheek sufficed for the moment I boarded the bus to the plane back to London. A weather-induced two-hour airport delay then accentuated my unease, which continued until the clouds broke away to reveal the lights of southern England. Soon I would be back in a country where I could feel at home.

When I finally got back to the Cavendish, a most disturbing letter was waiting on my desk. Dated November 8, 1955, it was from George Gamow, writing from the house in Bethesda that he soon was to sell. His decision to divorce Rho, and suffer its awful financial consequences, was now made. The letter opened innocently with a request for the Pasadena haberdasher's address so that he could order more RNA ties. Geo then asked what I thought about "Rundle's paper" and whether I thought he should be made an "honorable member" or whether it might be best to disband our club. My intestine already in spasms, I feared the worst of my scientific bad dreams had come true. Somehow the RNA structure had been solved by a chemist I didn't even know was in the race. To my despair, Rundle's model was apparently not boring, it explained a fact that I had never got to grips with—that all known proteins inexplicably are made up of $3N$ amino acids (9, 9, 21, 30, 39, 126, and also 135 in tobacco mosaic virus). Reporting that he was considering returning to cosmology, Geo ended "Oh gosh!"

Totally disconsolate by being knocked out of the ring, I waited a morose hour before Alex Rich uncharacteristically showed up in the lab before Francis Crick did. Seemingly equally discombobulated, Alex revealed that he, too, had received a letter from Geo that almost casually ended with mention of Rundle's home run. In fact, Alex earlier had news of the big breakthrough in a letter to Jane and him from Max Del- brück. Surprisingly neither letter explained what Rundle's brainstorm was. To learn what this clever inorganic chemist at Iowa State University had come up with, Max told Alex to get the latest issue of the *Journal of the American Chemical Society*. Hoping against hope that one of Cambridge's chemists might be paying for this prestigious journal to come by air mail, Francis and Alex hopped from one true chemist's office to another, only to find older issues that came by boat.

By then resigned to searching for another scientific goal, I became curious why none of the crystallographers back in Alex's lab at the National Institute of Health (NIH) had copied the Rundle article and sent it to us. Was it because they were too zombie-like in their shock? Alex had wondered likewise and finally had the courage to phone his lab. There he was put through to Jack Dunitz, who had the chemical know-how to set matters straight. Sensing that Jack was evasive, Alex could not keep Francis, standing next to him, from soon grabbing the phone. By then, both had smelled a hoax, and the mumble jumble coming back confirmed that our Cavendish group had been the victims of Gamow's latest practical joke.

Feeling as if coming off a hammer blow, I went that night to the Michaelmas term feast at Peterhouse, my engraved invitation saying "Doctors (Ph.D.s from Cambridge or Oxford) wear scarlet (robes)," indicating that the occasion was a "white-tie" affair. John Kendrew, fearing that I could not master such fancy dress, warned others that I might appear inappropriately. But with Francis also on the guest list, I went around to Portugal Place beforehand for Odile to check me over. With Peterhouse long proud of its High Table cuisine, I expected to be floored by its many courses, each with its special wine. The evening, however, proved—at least for me—less ebullient than the more understated black-tie feast at Christ's to which Hugh Huxley had invited me earlier.

Hugh was no longer embarrassed by the emotional triangle that he saw himself almost unfairly sucked into. It came as a relief to him, therefore, that the news of the Kendrews' irreversible separation had become semi-public. Those of us in the know felt that this mismatched couple were bound to part. The fact that John treated many college functions as more important than being at his Tennis Court Road home had to mean that his mind usually went in directions other than towards his wife. It was widely thought unfair that Hugh felt he had no choice but to move elsewhere. But whether he himself still wanted to be part of the Cambridge scene, was a question no one thought he yet wanted to be asked.

In the basement X-ray room, Don Caspar was finalizing his last experiments before writing up for *Nature* his evidence of fivefold symmetry in spherical plant viruses. While I was in Germany, he temporarily had awful fears that he might have accidentally exposed himself to a nasty, if not fatal, encounter with the high-intensity X-ray beam emitted from the powerful Cavendish rotating anode X-ray generator. Late at night, too tired to notice, Don neglected to put the appropriate heavy lead shielding between himself and the X-ray beam. All too soon realizing his mistake, he spent the next several days wretchedly scared of developing signs of incurable radiation-induced skin ulcers. Fortunately not even momentary redness appeared on his hands, and after a week Don was back taking X-ray pictures.

Much less terrifying was the verbal fight that he recently had had with Rosalind Franklin. When she came up to Cambridge for a day's visit, Don learned that Rosalind was collecting crystals of turnip yellow mosaic virus to put herself in competition with him. To him, this was a low blow because she and her research group had more good things to do with tobacco mosaic virus than time to achieve them. To my surprise I, not Francis, was left that afternoon to arbitrate between them. Using diplomatic charm that I never before possessed, I seemingly convinced Rosalind of the unfairness of her climbing up Don's back. Badly needing a strong drink after she had gone, I took Jane to the Bath Hotel bar to let her watch me calm down over two whiskies.

Soon after, I saw Victor Rothschild, who had sent me a recent message about Rosalind behaving like a hornet. The Agricultural Research Council (ARC) had just turned down her request for more funds, and, in reply, she face-to-face blew her stack at Sir William Slater, the ARC's director, implying he was not competent to judge her request. She had a point because he had been goose-like in suggesting that she travel 400 miles up to Aberdeen to use a totally inappropriate X-ray source that the ARC had given one of its grantees. I pleaded with Victor that Rosalind's sometimes awkward manner was more than compensated for by the importance of her research, as well as her intelligence and tenacity to succeed. I could not understand why Victor was now backing good form over brains in terms of the ways the ARC should use its resources. A year later, however, Rosalind was able to buy the needed diffractometer using American money from the NIH that Don helped her obtain.

Meanwhile, a diverting card from Linda Pauling let me know that she was up to her romantic ambiguities. In a letter to Odile Crick, she took pride in her ironing of eight shirts at the same time while complaining her parents were still beastly to her, with Ava Helen having higher expectations for her daughter than domestic duties. To me she cheerfully penned, "Comedy proceeding with hero and heroine playing parts well; latter pleased with her performance—hero charming—a great success but obviously not perfectly at ease with such a part. The talent is there, however, and just needs to be developed. At times heroine feels need of moral support but is coming through all right."

I still remained uneasy over my Munich visit but Christa reassuringly wrote that she would be joining me at Dick and Nou Mitchison's Scottish home over the New Year holiday. By then Alex Rich, having successfully turned a six-week into a six-month visit, knew that he had to be back at NIH. For the most part he and Francis were still entrapped modeling collagen in preparation for revealing their hands at a meeting with their King's London competitors some 10 days hence. RNA, however, was what Alex had come to Cambridge to throw fresh light on, and with the "Rundle hoax" still in our consciousness, we made one last effort, in early December, to ask what its X-ray pattern told us. Here, data from Severo Ochoa's synthetic RNA-like molecules dominated our thinking. We could stop worrying whether branches came off the 2′-hydroxyl groups because one enzyme would not be able to make two such different bonds. And very definitely the sugar phosphate backbone of RNA had to be on the outside of a helical molecule. The absence of strong equatorial X-ray reflections ruled out a structure in which the heavy phosphate groups were part of a dense central core.

Alex and I now saw the basic RNA configuration as that of a single-stranded helix, very similar in dimension to that of the single DNA chains in the double helix. What until now we hadn't appreciated was how close the chains interpenetrated each other. Now half-believing we finally had the right answer, Alex and I drafted a manuscript to be vetted by Francis when we got his intellect temporarily off collagen.

To take my mind off our likely boring answer for RNA, I began playing squash on the Clare courts. Although I could beat Hugh Huxley, Leslie Orgel could humiliate me by skillfully flicking his wrists to put most of his shots out of my reach. Light relief came from Geo Gamow sending me a copy of his letter to Cornelius Rhodes, the boss of the Sloan-Kettering Cancer Institute in New York. In it, Geo replied to Rhodes's criticism that he had not mentioned in his recent *Scientific American* article that some of the DNA used for the X-ray analysis had come from an Englishman working in his institute. The finding of the double helix, not isolating DNA, was what mattered, Geo said.

With the Michaelmas term ended, there was a rash of parties—one a farewell party for Jane at Francis and Odile's. She preferred to risk awful seasickness on the S.S. *America* rather than anxious flying over the Atlantic. "Doctors wore scarlet" at the very alcoholic Blythe Feast at Clare to which Michael Stoker invited me. Wearing tails by then almost seemed natural. On one side of me was Boris Ord, the Head of the King's College Choir School, talking about his pre-adolescent boys' miraculous soprano voices. A week later I saw him and his charges when Jane and I went to the chapel for Schütz's *Christmas Story*. The weather had turned almost balmy, and we sat in our pews not minding that the chapel's vast space was not heated.

By then John Kendrew's plans to be in New York over the holidays had fallen apart, and I agreed to go walking with him over Christmas in the Lake District, just south of the Scottish border. From there, Christa, John, and I would go on to the Mitchisons' at Carradale. The arrival of a

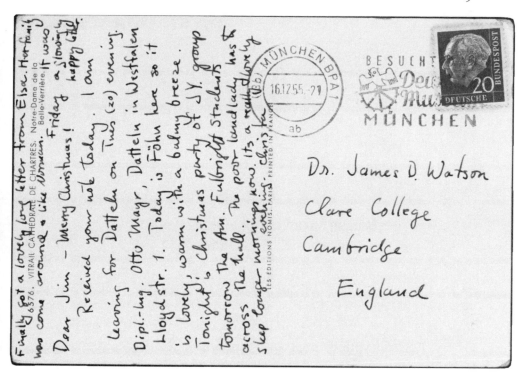

Finally got a lovely long letter from Elsa. Her family was come around & like Nomain. Friday a slowing happy & it was

Dear Jim — Merry Christmas! Received your note today. I am leaving for Datteln on Tues (20) evening. Dipl.-Ing. Otto Mayr, Datteln in Westfalen, Lloydstr. 1. Today is Föhn here so it is lovely, warm with a balmy breeze. Tonight's Christmas party of SY. group tomorrow the Am. Fulbright students across the hall. The poor landlady has to sleep longer mornings now. It's a really lovely evening. Chris

Dr. James D. Watson
Clare College
Cambridge
England

"Merry Christmas" postcard from Christa gave me the address in Westfalen to which I could write her about when and where she should join us. But whether Linda could greet us at Carradale was unclear. Av had not yet given her the word to come, seemingly not wanting to raise the issue by being ambiguous about where he would himself spend Christmas. Going to Carradale, however, had become a matter of pride to Linda. As his au-pair girl, she wanted more than vicariously to enjoy the privileges that go with being a Mitchison.

26

English Lake District and Scotland: December 1955–January 1956

ALTHOUGH THE ENGLISH Lake District is famous for its rain as well as its natural beauty, I did not know how heavy the odds are for getting soaked cold over Christmas. John Kendrew was taking me to the wettest place in England, with an excess of 130 inches of rainfall each year, some four times more than keeps the Cambridge colleges' lawns so breathtakingly green. The truck-filled Great North Road (A1) took us to the east of Leeds, before we cut west across the valleys of green and rock of the Yorkshire dales. On the way up, rain fell only intermittently, and I kept hoping that the clouds would part once we reached our Victorian-era hotel. But when we did arrive it was pouring buckets, and just getting out of the car so wetted my trousers that they remained damp through our evening meal in a barely warmed dining room more attuned to August days. In the summer, of course, it would be filled with light-hearted hikers or Wordsworth enthusiasts eager to see his Dove Cottage home in nearby Grasmere.

John knew well what we might endure and had brought along some heavy oilskin slickers, but they were not made to protect lower trousers or hiking boots. Initially our objective was to get to the top of the 3162-foot Scafell Pike, but even the bracken-surrounded lower paths were mushily unpleasant. And when they turned into steep, constantly crumbling scree, I questioned our sanity, knowing that all the other hotel guests were immersed in books inherently more satisfying than climbing towards objectives we were not likely to see through the blowing cold rain. Later, the Christmas roast beef fortified with red wine in no way compensated for BBC forecasts of more heavy rain coming off the Irish Sea over the next several days. No matter our increasing precautions, all our future ventures out into the blasting rain led to boots and clothes so soaked that there would be no chance, even in front of fireplaces, of their drying out before the next morning's return to the unmitigated hell of walking to get even wetter.

Christa's boat train was to get her into Victoria Station in time for her to catch the night train to Glasgow. Best for us would be for her to leave the train just before the Scottish border. I so telegraphed, hoping my message would reach her in time. Later, not sure whether she might have gone on to Glasgow, I happily found her awaiting us at the Carlisle station café. Exhausted by two nights of virtually no sleep, she nodded off as John drove across the hills of Galloway and through the gray-stoned grim of Glasgow. Only along the west shore of Loch Lomond did she come to life and talk about Christmas with her father's brother in Westfalen. Stopping for lunch before crossing the high moors on the way to Loch Fyne and the long ride along it, Christa seemed more eager to talk to John than me. But what so clearly upset her eluded me. Finally we were across the Crinan Canal with 15 sheep-dominated miles of driving finally bringing us to a large forestry plantation. From there we gently descended into Carradale.

Our entry that day to Carradale House was into the back through the larder into the kitchen, where afternoon tea was being prepared. Beyond the dining room, the large fireplace-dominated drawing room was filled with family and guests sprawled over chairs and couches reading books and newspapers. I soon spotted Linda, clearly happy to have me on hand for needed confidence-sharing. Having met Christa earlier in Cambridge, she volunteered to take her upstairs to where she would be sleeping while I sought out Nou Mitchison to learn the room in which I was to stay. Later, in the north-facing study filled with Wyndham Lewis drawings, Linda told me that she was not there because of Av's invitation. Instead his older brother, Murdoch, and his wife came to her rescue by asking her to be their guest for the holidays. Already here for several days, Av, with Linda temporarily no longer his charge, nonetheless remained elliptical about his needs. In contrast, his youngest sibling, Val, now had the security of being married well to someone intelligent as well as affable.

By catching Mark Arnold Foster, the labor correspondent for *The Guardian,* Val no longer felt it necessary verbally to put me in my proper place. Instead, she was almost affectionate, seeing that the world—at least tonight—was not going my way. After Christa had placed her bags in her room and had come down for some tea and scones, there was no hint from her eyes or in choosing where to sit that she was a close part of my life. Then, after the teapots were back in the kitchen and the main room now unoccupied with family and guests off in other places, she no longer avoided my face and blurted out what she had to say—that she was not at all in love with me and knew her mind and needs well enough to know that she would always feel this way. As hard as she tried over the past year, she found it impossible to convert her liking for me into the deep love needed to share her life completely with another person. Forcefully said, these were not off-the-cuff remarks but came as if repeated over and over in her mind ever since leaving Munich.

There was no opening for me to argue back. It was her feelings, not mine, that now mattered. Desperately that hour I wanted to believe that she would see things differently when she had had a good night's sleep. Deep down, however, I knew that in a more voiceless manner I had received the same message in Munich, but had not listened. Over dinner, Christa, not wanting to seem what she was not, conspicuously chose company at the other end of the long table.

Going later into the drawing room for coffee, I lay down in front of the fireplace wanting to be warm if I had to be unhappy. The intellectual word games, for which I was so unequipped four years before, were equally so now. They started as soon as the coffee was gone and there were no more chocolates to munch down. Dominating the games were Av's older sister, Lois, a product of Oxford's Lady Margaret Hall, and Val, a Somerville intellectual, who never willingly conceded that her Oxford-trained brain did not work as fast as her brother's. Also ferociously getting the answers right was Rowy, Murdoch's robustly handsome wife, already the author of a book on history—an expected accomplishment for a granddaughter of one of Balliol College's most illustrious masters. On one of her turns to answer, Christa popped up with an almost-right answer, while I might as well have been a zombie, given my inability to bring forth any answers either correct or amusingly wrong. Before tonight, the thought that I had a girl who was bright as well as pretty would be making me feel on top of the world. Now the prospect of losing brains as well as looks made me feel even sicker.

The next morning, Christa came down late after plans had been finalized for a longish walk that was likely to go beyond lunch. The sun was out, but rain was in other portions of the sky when John and I set off with Murdoch and Av for a long slog through the moors to the west of Lochgilphead. The ruin of an ancient castle was our objective, but when we got there it hardly seemed worth waddling across four miles of spongy, water-soaked heather. Apples and hard-crusted sandwiches of ham and cheese kept us going with darkness coming on just before we got back. Christa was pleasant, but no more so in seeing that I got tea and biscuits, with it plain that she and Linda had been talking out their lives while the big hike was on.

Not then easy to share confidences with was Nou's guest, the writer Doris Lessing, then openly in contempt of American culture. Two bad marriages and three small children on top of a British colonial upbringing, which started in Persia and ended in Rhodesia, made her easier to take through her writings than her black-haired, bristling personality. But her inherently full-bodied sexuality led one of Av's former Oxford companions to brag openly that she would be bedded by him before the holidays ended. No evidence later came that his late-afternoon attempts to so enliven Doris came even close to happening. Nor did Nou's attempt to have Doris temporarily share thoughts with me on long walks. After we were pushed off as part of a group towards a distant hill, I stayed silent for an hour. Seeing a need to explain myself, but not wanting to reveal my agonies at being jilted by Christa, I blurted out I could talk easily only to Francis. I immediately regretted my words, and was not surprised that Doris also stayed mute as we walked back through Forestry Commission lands.

At dinner that night, I sat next to Dick, then 65, and always curious about where the science that propelled his sons forward was going. As the still-large and powerfully built white-haired Labour MP from the Midlands town of Kettering, Dick saw his role more to satisfy his constituents' needs than his own ego. Back in the depression-ridden gloom of the early 1930s, Dick put out a little book called *The First Workers' Government* believing then that raw capitalism would never lead to a satisfactory society. As early as 1931 and again in 1935, Dick had tried for a seat in the House of Commons, but only when there was a Labour landslide in June 1945 did he get elected. It was their politics and the fact they so admired each other that kept them together so long, despite Nou's use of her celebrated novels to express passions that came to her so easily and strongly. Such attachments to others were never one-sided, with Dick's personal guest over this holiday being Tish, a willowy close friend of years ago.

Linda was not the person now to share the depths of my current Christa despair. Nor was Val, who I anticipated would tell me to seek a sensible girl who liked me much more than I thought I liked her. So I poured out my woes to Nou on the morning of my third Carradale day. She had personally long known the agony of loves misspent, and I was far from the first guest falling apart over a Carradale holiday. With Nou letting me monopolize our talk, I kept coming back to how easy it was to be with Christa. Given all the common feelings we shared about common acquaintances, particularly those from Cold Spring Harbor, it seemed to me that we would not later be torn apart by fundamentally different values. Of course, I worried whether she yearned more for an ebullient, German romantic type than my sort of unsentimental, pragmatic mind. There was also that awful German food, but we were both Americans and had grown up liking its hamburgers and hot dogs.

Not wanting to make Christa leave the drawing room if I was there also to read the London papers when they arrived in the afternoon, I gravitated to the play area for Murdoch and Rowy's two children, neither much past five but capable of precise speech far beyond their ages. With them I could use adult words, unencumbered by fear of their emotional overtones and yet enjoy the words that came back in response. In particular, Sally, like Victor Rothschild's young Emma, was years ahead of her age. Later, when Dick had started offering to make pre-dinner drinks, Nou waved me to come into her study saying that she and Christa had just finished a long heart-to-heart.

The news was not good. Christa was not going to trot, much less slide, back to me soon, if ever. She had told Nou that for more than a year she had felt trapped in an emotional box that was stifling her freedom. She wanted a bigger cage, if not wide open fields to explore. My jitters with our bodies worried her, and maybe she needed a more phlegmatic male, who would not take her so seriously and control her a little when she got out of hand. For my own good, Nou told me I should get Christa out of my mind. Almost unable to swallow, I knew that by wanting Christa so much I had never given myself a realistic chance of succeeding.

On New Year's Eve, everyone from Carradale House joined the local villagers at the local community hall for the fiddler-led dance that would go on to midnight. Long uncomfortable on a dance floor, and not at all equal to intricate highland reels, I first stayed on the side and looked towards Christa. When Nou got me on the floor, I only half successfully glided her much shorter and once-ripe body around the room. Av momentarily rose to the occasion of dancing with Linda but then ungallantly looked relieved when Murdoch cut in, not trusting Av to know how to disengage himself politely. Finally, when I had the courage to ask Christa to dance with me, she did not hold herself back, and fleetingly I remembered how we once happily square-danced at Cold Spring Harbor. Afterwards she was again very much part of the bigger group until midnight arrived, the dancing stopped, and, after the bagpipes sounded, we belted out "Auld Lang Syne." It was a time for kissing, and Linda and I let ourselves go as if we were more than we were. I had to kiss Christa, too, and she didn't resist. But then her face suddenly went blank, and we talked to others as we made our way along the path back to Carradale House.

New Year's Day was on a Sunday that year, and only the next day did the house party break up. John drove Christa and me to Glasgow Central Station, while he drove back to Cambridge. I could not bring myself to watch Christa go off to London alone, and so I waited with her for several hours until the overnight train arrived. Books and magazines marked the silence between us and kept me from too many last looks at her face. Later, after little sleep for several hours, I felt cold and blank in the Underground that took us around to Victoria Station. Walking up into its big drafty hall, the only consolation was that the boat train was still several hours from departing. But it was hard to put anything but despair on my face as we ate our cold, uninspired station breakfast of bacon, eggs, and baked beans.

The time to say good-bye was on hand. Even a slight hug, now, would be wrong, and my last memory of Christa was her sleepless face disappearing into the train compartment. Walking back along the platform, I felt like vomiting.

Cambridge (England): January–February 1956

IN CAMBRIDGE, ON my return from Scotland, somehow I had to be more than a melancholic relic of a lytic love obsession. Retreating into college to take stock of my life had no chance of working. My room was heated, but this was its only good point: it had not a trace of ancient charm. The only inspiration that I might get from looking out onto the drab brick tower of the National Socialist–like University Library would be the desire to jump off it. Back in the Cavendish Lab, I saw no point in hiding the fact that my personal life had collapsed over the holidays. But to the males about, the less said the better. Women-wise, Linda Pauling was still in Scotland and Mariette Robertson had not yet returned from her parents' temporary abode in Paris. Odile Crick, of course, might have recently spotted some newish girl's face with me in mind. But past experience suggested that such Cambridge popsies were inevitably already part of someone else's life.

I was not an animated sight over lab coffee. Now that Alex Rich had gone back to Bethesda, Francis was keen to finish our long-put-off manuscript on the structure of small viruses. But that morning in January 1956 my brain felt too heavy to have alternative thoughts about how we should phrase our arguments. More pressing was the wallet my parents had sent me for Christmas. Ordinarily I would have kept it in reserve because my current one, though shabby, was functional. Dumping it now could be the occasion for the picture of Christa I kept in it to go into the wastebasket, too. It had come with her birthday present to me of last

April, which had seemingly implied that she finally would be mine. The time was on hand to tear it up, so that I would not have a later opportunity to retrieve it from the trash dump. Peter Pauling, however, cautioned me to keep the snapshot intact. It's better to be able to look back at a true love, he said, than to recall girls that never made your heart stop.

Over tea, Ann Cullis happily was in no immediate hurry to get back to her work for Max Perutz. Sensing my abject demeanor, she volunteered to cook me supper at The Green Door. Just after seven, I took my shell-shocked psyche up its narrow staircase, reassured that Ann knew firsthand the warm personality of my now lost love. Quickly downing an offered sherry and the refill that immediately followed, I soon stopped feeling sorry about my fate. My hostess's strawberry blond mane and cooking skills made my mind instead turn to the unwanted fact that Ann's weekends were dominated by an admirer, who came up from London in a vulgarly large and fast car. Yet I momentarily thought she might, soon, look at me as more than simply Christa-less. But as I left her flat, Ann matter-of-factly wished me good luck in finding someone else.

The next morning, I saw the need to pen a note to Ernst but two weeks were to pass before I could compose a letter worth sending.

> Cavendish Laboratory
> Free School Lane
> Cambridge
> January 21, 1955 [=1956]

Dear Ernst

I imagine that Christa has written to you that she has decided not to marry me. As you might expect, this depresses more than a little, both because I love her and even more because I have always and even now think that we are basically very well suited to each other. Her answer seemed ambiguous and certainly her mood did not waver during the week at Carradale. So I imagine that I should take her answer as final. But this is very difficult, both for emotional reasons, and also because of the complete change from her September mood (and in letters afterwards) when I know she was quite in love with me and indicated that she would marry me. In Munich in

November, the atmosphere was tense, partly because I just didn't fit into her student existence and also because she was more than a little annoyed at my insistence that she see a first rate heart specialist. However, I didn't much think about it and hoped that the Christmas visit might be more pleasant.

As you no doubt were aware, Christa's attitude toward me has always oscillated between extreme affection and almost complete indifference and so I cannot be sure that this is not one of her periodic phases and that she might view me differently on a new occasion. In a way, I think I understand her very well and I suspected that she would break things off before she could ever decide how she felt about me. At times she impresses me as still very young and not really knowing what she wants except for the ability to grow up and choose her own life. The age difference between us was very apparent in Munich and it was obvious that she preferred to be with young people who had their lives completely in front of them. She was constantly irritated by the fact that I didn't lead the life of a penniless German student and instead preferred to travel by plane, eat good food, sit down at concerts, and generally adopt attitudes common to the English Middle Class among whom I now live. However I know very well that in about 3 years her attitude toward these things will radically change once these attributes of mine will no longer seem so disquicting.

The basic reason she gave for her decision was that she didn't love me any more and I know very well that she shall never marry anyone with whom she is not in love. But I also know that her ideas about love are very simple and that she shall find it a far more complex emotion than she now believes. Until I was 25, I was incapable of complete love and so Christa's attitude is no surprise to me. And so I cannot be sure that her decision reflects an inherent inability to marry me or whether it arises from a feeling created largely by a situation thrust upon her before she was ready. I was always aware that the matter shouldn't be forced but this was difficult to achieve since at times she was obviously in love with me.

Naturally there is no point in trying to see her again before she returns to the States. Probably for the first time, she now feels completely free to act as she pleases and so it should be much easier for her to see if she finds anyone with whom she is basically compatible. For my part I cannot exist under the hope that Christa shall eventually marry me and so if I were to meet another girl that I could talk to as easily I should naturally try to marry her. The bloody trouble is that Christa is very much part of my system and I am afraid that it may take a long time to remove her from my dreams, as well as my reality.

In a way, I would like to hear your views on the matter, largely for the reason that you know Christa very well. But I can also understand why you might prefer to say nothing.

Otherwise my fortunes fare well. There is a good chance that arrangement of RNA within TMV shall be known before I return in early June and likewise the structure of Poly Adenylic Acid looks hopeful. Over the Easter holiday, I shall be lecturing in Israel and I shall call in at Cairo if Jeffries Wyman is about at this time.

With my usual best greetings to Gretel and Suzie

Jim

Don Caspar, now back from Christmas skiing in Austria, gave me the fresh news that Rosalind Franklin's group at Birkbeck College London had just located the RNA within tobacco mosaic virus (TMV). They did this by comparing the X-ray diffraction pattern of TMV rods with those from similarly shaped RNA-free rods, made by repolymerizing pure TMV-protein subunits sent from Tübingen. Most excitingly the very prominent TMV density maximum at a radius of 40 angstroms was absent in the particles lacking nucleic acid. Because the phosphate groups in RNA effectively act as "heavy atoms" in diffracting X-rays, the 40-angstrom density maximum had to represent the sugar-phosphate backbone of its RNA component. So RNA's structure within TMV had to be completely different from that assumed when it folds up free in the absence of protein. By then our office no longer had space for more model-building following the arrival of Joe Kraut, John Kendrew's newest American postdoc, here to work on myoglobin. At times, there were seven of us jammed together in our office, all trying to do our own thing. So I began daydreaming of a potential theoretical note to *Nature* by

Francis, Aaron, and Joe. The alliterative Crick, Kraut, and Klug assemblance would make Gamow jealous.

Impatient to publish her new TMV finding quickly, Rosalind was in a quandary as to how to give credit to Don's earlier, unpublished Yale work. By now, their past squabbling about who should explore fivefold virus symmetry was a memory much better repressed. Rosalind, knowing of Don's writer's block, was actually writing up his Yale results so that she could properly acknowledge it in a companion manuscript to be submitted to *Nature*.

> Birkbeck College Crystallography Laboratory
> (University of London)
> 21 Torrington Square, W.C.1
> Feb 10 [1956]

Dear Jim

Here is a very hasty version of Don's note—I have done no rewriting, as you and Don will obviously want to do some anyway. I will send mine on as soon as it is finished (probably Monday). I'm sending a copy of this separately to Don in case it doesn't reach you before you go to Scotland.

> Yours
> Rosalind

Spurred on by Rosalind's help, Don himself was writing up his observations on bushy stunt virus. Seeing great value in having Francis's and my theoretical article on "Structure of small viruses" appear next to it in *Nature,* I stopped procrastinating and got on to its writing. A CIBA Foundation meeting on "The Nature of Viruses" was to be held in London in late March and our papers should be out before. So motivated, the manuscript was soon finished and submitted to *Nature* on January 23.

By then, Linda was back. Remaining in a large, cold Edinburgh flat was no longer a sensible option, once she saw that Av never intended to give her the au-pair girl's option of rebuffing advances from the household head. Moreover, the third of their ménage, John Godfrey, had started to make unmistakable hints that he was testosterone-driven to chase her. At best, Linda found this embarrassing, for it was Av's bashful Haldane eccentricity, not his devotion to laboratory animals, that had led her to Edinburgh. Accepting advances from an underpaid socialist, obsessed by voles, wasn't part of her strategy. On the morning of her departure, Av's gallantry reappeared as he gave her the latest issues of *Vogue* and *The New Statesman* to read on the train. Linda now could take comfort in a unique experience, in which she had surmounted adversity, displayed much grace, and avoided chilblains.

Mariette was also now back in town and happily spending more time with Peter, who had stopped driving out to Girton. Apparently Julia Lewis, his flame of the fall, had a bad flu bug that would not go away and she had not come back for the winter term's first week of lectures. Linda and I were together more and more often, going to movies at the Arts Theatre and seeking out Alfred Tissières, whose rooms' high ceilings and views out onto the Chapel invariably made having tea or evening coffee there a morale-sustaining event. Although the North Sea wind blew cold occasionally, there were enough almost balmy days above 50°F to let us take long walks along the Cam, sometimes as far as Granchester.

Also calming my nerves was the requirement that I soon prepare a manuscript for the April meeting on macromolecules in Israel. Between such writing moments, Crick, Orgel, and I kept coming back to RNA's role in protein synthesis. Francis would not give up his bizarre idea of a year ago that small RNA adaptor molecules were somehow involved in reading the genetic code. Leslie and I, however, continued to think this scheme was too off-the-wall to consider unless physical evidence for RNA adaptors was somehow found. But we had to admit that we saw no obvious way to generate hydrophobic cavities along the surfaces of stacked bases that could distinguish between the amino acids leucine, isoleucine, and valine.

That Linda and I were now so often together having afternoon tea at The Whim or Copper Kettle did not go unnoticed. To my surprise, I learned from Alfred that rumors were flying that a Watson-Pauling affair was on. When I told Linda about our "affair," she was instantly pleased and amused by not having returned to Cambridge unnoticed, seeing no

*Mariette Robertson punting on the River Cam in
Cambridge, spring 1956*

reason to deny others the satisfaction of their gossip. Soon we were onto concocting a plan for escalating the rumor by jointly hosting a big, stylish evening party. Those unexpectedly receiving such invitations at short notice would be bound to wonder whether we were using the occasion to announce our engagement.

More than a week passed before we finally moved into action. Our belief that only Alfred's tall-ceilinged rooms at King's would be grand enough was complicated by the fact that, several days before, Alfred had left for three weeks in Amsterdam to follow up experiments started before a friend moved there to a professorship. Believing, however, that he would not be unduly upset, we soon got our hands on some 100 appropriate stiff white cards. On them, Linda stylishly inscribed:

Jim Watson and Linda Pauling
Invite You to a Party
In Alfred Tissières' Gibbs Building Rooms
King's College
Saturday, 18 February 1956
9 p.m. onward

Anxious to get at least several with-it, unattached girls to give our party panache as well as academic respectability, Linda had popped out to Girton. She couldn't dig up Julia but did find her close friend Janet Stewart, who, inexplicably, did not know when Julia would return. Fortunately Janet promised to make an appearance and bring with her several lively friends to make the evening one in which Francis would enjoy displaying his conversational skills. Although we had invited all the academic bigwigs who might know who Linda and I were, we remained uncertain until the last moment how many would show. There was also the question of whether Alfred actually knew how big the party in his rooms would be. Frightened by the last-minute possibility of not being able to get into locked rooms, Don Caspar got reassurance from higher-ups at King's that they would not blow the whistle.

Not unexpectedly, the first guest to arrive precisely at nine was Freddie Gutfreund, always eager to talk enzyme kinetics that I never wanted to understand. Soon we had the first married couple when Roy Markham and wife Margaret entered, coming early enough to insure that they might not be victims of our incompetence in not buying enough wine. Some 30 guests were on hand half an hour later when Cambridge's big chemist, Alex Todd, arrived with his wife Allison. Linda and I, knowing the moment of the evening had arrived, fondly put our arms around each other to give the Todds the message that the party was what they thought it was for. Standing upright, inches above the rest of us, Alex personally knew what it was like to have a Nobel Prize–winning father-in-law. His wife's father, Sir Henry Dale, had discovered how acetylcholine carries signals from nerve cells to muscle cells. Worried that the Todds might quickly not find the occasion worth many minutes, I began to relax when they stayed contented for at least 45 minutes. Another sign of the party's success was the arrival of

Alfred's friends, Noel and Gabby Annan. Not only was Noel to be the new Provost at King's, he had already attracted much attention from his recent biography of Sir Leslie Stephen, Virginia Woolf and Vanessa Bell's father. His and Gabby's ebullient presence gave the evening its necessary King's verve.

Linda had hopes that Victor Rothschild might come by briefly and, gallantly, flirt. But from Peter we learned that Victor and Tess were otherwise engaged. Earlier, Peter had gone to Merton Hall to borrow a top hat and opera cloak dating from Victor's less full-bodied youth. There he bumped into Sarah, a daughter from Victor's first marriage, and told her that accepting our invitation would be worth the trouble of dressing up. But there seemed little chance that she would join us now. Before Peter walked in with Mariette on his arm, neither Linda nor I knew whom he would come with. Knowing that Mariette would not like to come alone to face her rival, Leslie and Alice were to bring her if that proved necessary. But Julia was not part of the evening; Peter's quiet, relaxed grip on Mariette's hand identified a couple accepting each other. Handsome in black tie, Peter's often outpouring of verbal charm was absent, replaced by almost sentimental concern about where he and his Cambridge friends' lives were going.

When the Girton contingent came in en masse led by Janet, Peter, with Mariette at his side, had the sense not to throw his arms around them. Instead, he pulled aside the male friend of Janet's, someone he obviously knew, who was reading international law. I moved towards Janet's elegantly draped fullness, hoping that her evening ahead might still have a measure of uncertainty. But much too fast, Gidon Gottlieb, her long blond-haired companion, moved to join us. Sensing that I was not the aesthete whose ideas move easily between French and English, I was only mildly annoyed some minutes later when John Kendrew took me aside to say that I was needed to bring Don Caspar back to life.

Don was now dead drunk from too many refills of red wine poured down to compensate for the continuing strain of not finding an attractive woman to share his thoughts with. Leaving John and Gidon to look at Janet's intelligently filled gown, I found Don on the floor of the wine-bottle-strewn larder. Quickly, I brought him outside to the grass. There, he lost more of his evening's imbibings, an earlier whack of which had not yet adequately been removed from the hallway floor. So relieved, Don determinedly groped his way upstairs, seeking out a beer chaser that he assured me would get him back into shape. Then he promised to start clearing away the abandoned wine glasses now in evidence as midnight approached and guests began leaving.

With the party now thinned enough for easy conversation, I gravitated with Linda back to Janet and Gidon, whose father, he let drop, was the agent in Paris for the French Rothschild family's philanthropy in Israel. I was also using beer to give my stomach a less-empty feeling with Linda achieving this same end with glasses of cold water. Relieved that the party had definitely clicked and that later postmortems would not embarrass us, we turned our attention to the mystery of Julia's absence. Peter was not one who easily lost hold of a pretty girl. So we began to suspect that Julia had not even briefly been back in Girton this term. With Janet possibly knowing more than she wanted to reveal past midnight, I suggested that I come out to Girton after Sunday lunch was over. Janet's serious smile let me know I might be needed.

Cambridge (England): February 1956

THE NEXT AFTERNOON, on Sunday, February 19, I cycled out to Girton, anticipating the astutely regal Janet Stewart and her explanation of Julia Lewis's disappearance. I hoped there might be less between her and Gidon Gottlieb than was conveyed by the way he made last night's shared thoughts into virtually beatific moments. Momentarily disappointed by finding him again at her side, I soon had to admit he had reason to be there. Julia was indeed in bad trouble. Janet immediately confessed that last night she had been holding back vital information. The party was not the place for a bombshell, which, as late as last night, she hoped need not explode. Several weeks before, Julia had let her know she was not at Girton because she was several months' pregnant. Morning sickness had hit, and her shape would soon publicly reveal the same message. Fearing from the start that Peter was the father, my stomach sank when Janet went on to say that Julia had been with no one else over the fall.

When Peter had learned he was to be a father, he clearly panicked. This was not the right time for him to get married. To start with, he could not afford a wife and child. All his current monies as a research student came from his parents, and he did not want to think about getting more under these circumstances. More to the point, even if he had his Ph.D. and was earning money, Julia was not the girl he wanted to jump into marriage with. He could never be as emotionally close to her as he once had been to the petite, blond au pair Nina. And he was still fond of Mariette Robertson—though not as much as she was of him. Moreover, he knew that he could not stop chasing new girls, to whom he would put on his Pauling charm and see where it led.

Immediately he heard of the news, Peter told Julia not to have the baby, offering to find the appropriate medical help. But Julia would hear nothing of this, despite several long telephone calls to her by an increasingly desperate Peter. With her holiday in Paris over, Mariette was by now on his side whenever he beckoned. Unlike Julia, who almost knew him best from excursions in the Mercedes roadster, Mariette had seen his moods go up and down many times without apparent reason. And her affection did not arise from unrealistic expectations as to what he or his family was like.

On the night of the party, Janet still thought the worst might not happen, but a new phone call this morning had just told that nothing she or I might now do could undo what had already passed. Julia's family, from Christmas on, knowing of her condition, was increasingly desperate to make Peter accept his fate. Several days earlier Julia's brother had come to Girton and told its Mistress why Julia had not returned for classes. She, in turn, scheduled an emergency appointment with the Master of Peterhouse, the college to which Peter was attached and that had let him live in the hostel across Trumpington Road from its main buildings.

Succinctly apprised of what Peter had not done, Peterhouse's Master, not seeing the need for further consultation, sent notice to Peter that he should immediately come to his study. There Peter was told that he was being sent down (expelled) and should leave his college rooms as soon as possible. If Peter had made Julia an honest woman earlier, the Master said, the embarrassment of the situation might have been glossed over, and as a married student Peter would have been allowed to remain in Cambridge. Now his Cavendish days were over and Peter—and most likely Mariette as well—knew this when they made their striking entrance into Alfred Tissières's rooms for Linda's and my party. Knowing that he was to "go down" forever, Peter wanted to be remembered for a last display of bravado. This was why he had gone to Merton Hall and told Victor Rothschild cryptically that the occasion demanded that he be seen in the 1930s dress appropriate for Victor's first days as the 27-year-old third Baron Rothschild.

Returning from Girton, feeling sick from my new knowledge and fearful of what was to happen next, I knew that even if Peter had entrusted me earlier with his disquieting news, its tragic course would have still followed. No one could have convinced Julia that charm and family fame do not by themselves make a good marriage. Although Julia had to know that the massive Mercedes and the Porsche were soon to go across the Atlantic as toys for a rich wife-entitled brother, Peter nonetheless had given her life zing. With him, she felt more than a scholarly and fragilely pretty daughter of an upright Midlands family, dedicated to hard work and respectable behavior.

As a Peterhouse Fellow, John Kendrew knew what was up by the time of our party but could spill the beans only when we saw each other just before Monday morning coffee. There had been nothing John could do to slow down the brusque way in which Peter was exiled from Peterhouse, and hence the university. Even if Peter's past three and a half years at Peterhouse had been paragons of virtuous behavior, the gravity of his current indiscretion left them no other course. Peter had been in and out of trouble for more minor infractions than the various college tutors wanted to remember. While they couldn't deny their American was likable, and added much to the college's general ambiance, no one was surprised that Peter had not instinctively done the right thing by his Girton girl. The time had come for purging from Peterhouse's midst its latest unfortunate example of youth hell-bent on pleasures of the flesh.

Peter now had to react to two crises, not one. Besides facing up to what to do about Julia, there was now the problem of what to do about his career. John came to his rescue by saying that he thought Peter might be able to transfer to the Royal Institution (RI) in London, with John continuing as his supervisor. Sir Lawrence Bragg had unsuccessfully urged John to move there when he became its director and would likely see Peter's arrival as a way to bring John's intellect into greater contact with the RI. And although Sir Lawrence superficially looked unbending, deep down he was compassionate and would not like to see Linus's son's life ruined.

Bragg's agreement to Peter's transference to the RI was conditional upon Peter making Julia an honest woman as quickly as possible. In fact, Peter needed no such threats once it was clear that a baby, borne as much from his loins as Julia's, was going to be born. During all of January until the awful Friday, February 17, Peter kept hoping against increasing reason that Julia would accept the lunacy of pinning her and a child's future on an up-and-down charmer able to find some attractive feature in almost any woman he came close to.

Linda, though initially angry with Peter for not taking precautions when he was with Julia, saw no alternative to a quickly arranged civil wedding witnessed by as few family and friends as possible. Realistically, like most of us close to Peter, she did not have high hopes for what would happen afterwards. But there was always the possibility that Peter would become more responsible after some of his freedom was taken away by the give-and-take restrictions of married life. Mariette, by then totally distraught, saw no happiness ahead for either Peter or herself. Staying alone at the Orgels' flat looking after their baby became an unbearable ordeal. Mariette knew Peter too well and too lovingly ever to have shotgunned him like Julia had done. But she had to reflect that Peter was virtually predestined for the shotgun. Would he have ever willingly commited himself to a monogamous institution that he was inherently unsuited for? To let out steam, Mariette biked back to central Cambridge whenever Alice was home. There she found an ever-decreasing number of acquaintances who could bear any more Pauling family gossip. But knowing how hard it was still for me to get Christa out of my mind, I was always there to listen.

The now-inevitable civil marriage ceremony took place in Cambridge's Register Office on Castle Hill. Julia, dressed well but not in white, was given away by her father. John took the best man's role hoping to show that Peter was still valued by those who mattered. My role was to take Mariette out to lunch—to make sure that she did not ferret out the wedding site and make a scene. After the marriage there was a small luncheon, dominated by Julia's immediate family.

In retrospect, Peter and Julia should have gone off quietly, but John thought a small evening party at his Tennis Court Road house might let the occasion end less inherently sad. Its outcome, however, was just the opposite. Word that free, copious champagne might flow from 9 p.m. onwards led to a continuous influx of uninvited spivvish drinking companions from Peter's past party moments. They soon outnumbered those with faces aware of what had happened today. John, possibly wishing to black out the origins of the evening, became unabashedly woozy and increasingly amorous toward Linda, herself not sure of the role she should be playing. Quite definitely she was not a bridesmaid.

All too soon, John's alcohol supply was gone, leading to a general exodus in search elsewhere for more of the drink that had made Tennis Court Road the evening's place to be. With the small drawing room now quiet enough for actual conversation, no one knew what was left to say. The time had come for Peter and Julia to leave. Peter put on a broad smile and Julia a harassed, uncertain one and to quiet applause they went out to the Porsche that he had parked outside.

29

Cambridge (England), Israel, and Egypt: March–April 1956

LINDA PAULING'S LIFE, in a much less dramatic way, also became subject to someone else's whim. With no apparent purpose, no job, no school paper to write, nor boyfriend to come to grips with, she admitted being at loose ends. But now she very much hoped that the Home Office did not see her this way. To her shock, one of its civil servants had given her a summons to appear before the Magistrates' Court on a charge of being an illegal alien. At first she couldn't believe that she had done anything wrong. Lots of her Cambridge American acquaintances were using small bits of parental money to let them pursue minor academic objectives and feel more special than their compatriots living stateside in Eisenhower's America. No one had told them to go home.

The American way of smiling away transgressions of silly laws was not now to work for Linda. To her horror, she found that English laws were there to be obeyed. At first, she was reassured that most Cambridge Justices of the Peace were prominent Cambridge women, often the wives of senior professors. So she was not prepared for the clipped voice of early-sixtyish Lady Adrian telling her that ignorance of the laws was never an excuse. A college graduate, she must have read the block of printed words in her passport stating the conditions under which she could stay in the U.K. and the length of time she could do so. Linda's halting answer—that she thought she had a month more before she had to go to the police station on St. Andrew's Street to renew her permit— got her nowhere. When they asked how she might finance a further stay in Cambridge, she admitted dependence on her parents who sent bank drafts to cover her needs. And asked when and for how much their last draft was, Linda knew there was trouble ahead. Getting a letter from her parents pledging further money might take more than time to get, and even such help might not do the trick. Without pausing to reflect on her next words, Lady Adrian fined Linda £5 and gave her two weeks to leave England.

Linda immediately smelled an anti-Pauling conspiracy. Lady Adrian, as the wife of Lord Adrian, the famous physiologist and Master of Trinity College, had to know of her father. It was not right for the wife of one Nobel Prize winner to boot out of England the well-intentioned daughter of another world-famous visitor to Stockholm. An appeal was possible, but Linda would have to act fast. It was not clear at all whether Lady Adrian knew who Linda's father was. And, if so, had Peter's ignominy reached the Master's Lodge at Trinity? Although Lady Adrian's no-nonsense manner was not that of a busy gossip, her young physiologist son and heir to the title, Richard, might have briefed his parents. If Peter's debacle had not been so immediate, Francis Crick or John Kendrew might have quietly gone to Richard to ask his mother to reverse course. But for Linda's friends at this time, working to keep her in England when everyone's sanity demanded that the Pauling name be off their lips as soon as possible, proved a hard task.

Going back to Pasadena at this moment was not what Linda wanted and being under her mother's thumb was the last thing she needed. There would be endless postmortems on how Peter got into his current situation and how they could have better handled his girl craziness. Even worse, her mother had just written to say that she had found Linda the perfect young Caltech faculty member. Barclay Kamb, according to Ava Helen, was both bright and nice and bound soon to be a tenured geophysicist at Caltech. The thought of her mother arranging her future instantly made Linda red and she decided instead to listen to John Kendrew's suggestion that his art-historian mother help her. A resident in Florence since John's boyhood, she had long been the mistress of a noted art authority, and a potentially perfect mentor to help Linda get a firsthand view of Florentine life and culture. With luck, a positive answer would come from Italy before she was kicked out of England.

Nervously Linda wrote her parents that she needed money to sidetrack to the artistic wonders of Florence: once on the continent, she could reassess her options.

To my chagrin, I also got a summons to the Magistrates' Court at the Guildhall. Learning of Linda's residency problem, I looked at my passport to discover that my permit had expired the week before. Going immediately to the police station, I hoped that my voluntary appearance would let my permit be routinely extended. Instead I was soon also in front of Lady Adrian, who came into the courtroom with a cane to handle her slight limp. After learning that I had ample money, lived in Clare, and worked at the Cavendish, my permit was extended until July 31. But I also was fined £5.

When I had almost finished revising the manuscript needed for my forthcoming early April visit to Israel, I received an exciting letter from Alfred Gierer. From my November Tübingen visit, I knew he was studying the purified tobacco mosaic virus (TMV) RNA component and so I had written him immediately when I learned of Rosalind Franklin's result that placed its sugar-phosphate backbone 40 angstroms out from the water-filled center of the virus. Now he revealed that his isolated RNA was infectious, not requiring TMV protein to aggregate around it for successful initiation of virus infection. This was a much cleaner result than Heinz Fraenkel-Conrat's last June report that protein as well as RNA was needed for infectivity.

Who was right might come out in London at the upcoming CIBA Foundation meeting on "The Biophysics and Biochemistry of Viruses" to which two persons from Berkeley's TMV contingent, Robley Williams and Art Knight, were coming. Gierer could meet them there, but my last-minute efforts for getting him from Germany went nowhere. Meanwhile, I went into London to get an Egyptian visa stamped on my passport so that I could visit Jeffries Wyman in Cairo after my 10-day stay in Israel. Going to Egypt after being in Israel was said to be tricky, but I was told that, if asked, the Israelis would see that my passport bore no sign of my being with them.

A preliminary program forewarned us that animal RNA viruses would be the main focus at this CIBA meeting to which Francis, Don Caspar, and I as well as Rosalind and Aaron Klug from Birkbeck College had been invited. Like all gatherings at CIBA's posh Regency House in Portland Place, attendance was limited to 35 persons. I knew just enough about most speakers to fear the worst from almost half of them. Many virologists still did not think in terms of genetic information and were wedded to sloppy immunological and biochemical procedures that would never go to the heart of what viruses were.

Spotting Robley Williams at the pre-meeting dinner on March 28, I immediately told him that Alfred Gierer was finding pure TMV RNA infectious. In response, Robley told that they, too, recently had some evidence that RNA preparations seemingly lacking any intact TMV rods could initiate viral infections. Even more important, he reported that when they reconstituted TMV rods from protein and RNA components coming from different TMV strains, the infections that resulted always displayed the disease symptom of the RNA partner, never that of the protein. To the amazement of André Lwoff, also drinking sherry with us, Robley seemed unaware of the bombshell implications of the combined Tübingen–Berkeley experimental results. The flatness of his tone made us wonder whether he had still not begun to think of RNA molecules as linear sources of genetic information.

Mischievously, André and I on CIBA Foundation stationery wrote out the words "TMV PROTEIN INFECTIOUS—BE CAUTIOUS—WENDELL" and had it passed to Robley the following morning as an overnight telegraph message coming from the Berkeley Virus Lab's Director. During his talk, given immediately after Francis's delivery of our joint paper, Robley continued to downplay his new results, emphasizing their preliminary nature. Later over afternoon tea, our curiosity led André and me to ask Robley his reaction to Wendell Stanley's message, saying that the telegram had mistakenly been first given to us. Unable to maintain straight faces, we admitted being the originators of a phoney message. Robley in reply told us he had questioned whether the message had been a hoax perpetuated by someone in Berkeley.

In Cambridge, by then, the steady tears and "what ifs" of the past tumultuous month had largely subsided. Linda was off to Florence to live in a pension chosen by John Kendrew's mother while Don and I kept Mariette Robertson company, each of us trying to show cheerfulness when we felt otherwise. Fortunately the brutal cold of February had passed and thousands of crocuses were up along the college backs. Three weeks to the date of Peter's nuptials, and just before the CIBA meeting started, Mariette injected some new faces into our lives by holding a small Saturday-night party at the Orgels'. There I met the young American physicist, Wally Gilbert, then supervised by the almost-as-youthful Pakistani Abdus Salam, who later set up the International Centre of Theoretical Physics in Trieste and won a Nobel Prize in 1979. Earlier Wally and I saw each other at a gathering of physicists listening to the mathematician and pioneer of computing Alan Turing. Already legendary for helping crack the top-secret German Enigma codes, his talk that night was on the almost mathematician-designed morpho-genetic patterns of plants, a feature noted by the English biologist D'Arcy Wentworth Thompson several decades before. The math of Tur-ing's argument was above me, but I comforted myself suspecting that it was unlikely to help biologists understand how growth patterns develop.

Also at the Orgel party was Wally's wife Celia, diminutive in size but not in spirit. They married when she was about to finish majoring in English at Smith College. More person-oriented than Wally, I told her that my ambition now was to find a rich wife. It was my response to her when she asked me what I could do to top the double helix. Celia imme-diately saw my major challenge ahead—scientists don't usually mix with people with money. But as I had so abjectly failed to win the daughter of a poor academic, I thought I might less upset a rich girl to whom my newly acquired Cambridge mannerisms would not seem foreign.

On my way to Israel on March 31, I stopped off in Athens, expecting spring warmth but finding temperatures only up to the mid-50s Fahren-heit. Alone and girlfriendless on top of the windy Acropolis, I had little inclination to delve into guidebook facts. It was a relief to go on to Israel for my weeklong meeting, finding hosts who knew who I was. At Tel Aviv Airport, I was shepherded with other new arrivals into cars to take us to a pleasant beach hotel. From there we were bused every morning to Rehovot, southeast of Tel Aviv, to the Weizmann Institute of Science. The first half of the "International Symposium on Macromolecules" occurred there before concluding in Jerusalem.

I happily had most of my meals with Weizmann Institute biologists, who correctly suspected my potential for being bored by too much chemistry. One night the microbiologist Ben Volcani brought me to his family's flat on the institute grounds where I could smell the nearby orange groves. There was a palpable fear in the evening air of impending war between Israel and its Arab neighbors and institute scientists were rotating guard duty at night to intercept infiltrators coming across from the nearby border of Jordan. The next night I was taken by Leo Sachs and Matilda Danon, who worked together on cancer cells, to an Arab dinner in the ancient Arab part of Tel Aviv.

The next day we avoided a chemist-filled excursion bus by going in Leo's just-runnable big car to Nazareth and then on to the Sea of Galilee, above which rose the Golan Heights and the Syrian guns that too often rained artillery shots onto Jewish farms below. On Sunday I moved to Jerusalem with Ephraim Katchalsky, the intellectually spirited Israeli protein chemist, who drove scarily fast up the winding highway into the Judean Hills. There I learned that it might be possible for me to go to Egypt via Jordan rather than backtracking through Cyprus. Tourists were allowed by Jordan to pass through the Mandelbaum Gate into the old part of Jerusalem, provided they could show they were not Jews. All I had to do was to obtain a certificate of church membership. So, after the Monday morning session at which I spoke, I went to the West Jerusalem YMCA. There one of the staff used a sheet of their stationery to write that in 1928, in Chicago, I had been baptized an Episcopalian at the Church of the Redeemer, my father's family church. This seemed not the moment to reveal I was a long-lapsed Catholic.

Armed with my fake certificate and my passport, my border crossing was routine and I found myself close to the walled old city. While walking towards it, I was pestered by youths wanting to be my guides. I tried unsuccessfully to move faster than they, but soon was surrounded by beggars of all ages wanting baksheesh. Immobilized, I settled on a fee with one of the younger teenaged boys and quickly he alone was heading me in the direction of the Al-Aqsa Mosque. My overnight stay was at a small, inexpensive hotel for Christians, right in the heart of an extraordinarily wondrous city. The next day, my guide and I went out to Bethlehem and the following afternoon, after a morning of buying trinkets, I was on a two-engine wartime C40 transport that let me look down at the great pyramids of Giza before landing at Cairo Airport.

I was met by Jeffries Wyman, whose driver took us to a small pension near the flat that Jeffries and his Russian-born wife, Olga, inhabited. The daughter of a Grand Duke, who once ruled the Crimea, Olga and her family had fled to Paris during the revolution, and it was there that Jeffries married her in 1954. She had a son from a previous marriage. At dinner, I was surprised to find her more like a strong-willed peasant than an aristocrat. The next day, I went with Jeffries to lunch at the Gezira Sporting Club, located on an island in the Nile. There we saw the Russian ambassador and his entourage, now suddenly a big factor in Egyptian life. When Gamal Abdel Nasser began buying arms from Soviet countries, John Foster Dulles cut off American funding for the proposed High Nile Dam at Aswan. So Nasser nationalized the Suez Canal to find the monies needed for the dam and chose Russian engineers to complete the project.

That afternoon, Jeffries took me out onto the Nile in a small sailboat and later put me on a night sleeper to Luxor, some 600 miles to the south. Telling me this would be the high point of my trip to Egypt, he was more than right. After lunch at my one-star hotel next to the much more opulent, river-facing Winter Palace Hotel, I made the two-mile walk down to the great temple of Karnak, whose vastness and unbelievably massive columns I was totally unprepared for. Roaming about it until sundown, my walk back was punctuated by prayers coming from the mud houses signifying the day-long Ramadan fast was over and feasting could begin. A simple ferry brought me the next morning across the Nile, where I was met by several small boys offering me their donkeys to ride to the Valley of the Kings and its fabled Tomb of Tutankhamen. I chose the one who offered the lowest price, but he had to beat his donkey continually to keep it going. My donkey ride ended at the 70-foot-high Colossi of Memnon, a pair of towering statues that were already a tourist attraction in Roman times. Here my donkey-beating guide wanted to be paid five times more than the fee he quoted to get my business. Angered by being tricked, I only doubled the original figure. But as the boy walked sullenly away, I regretted paying him less than a half dollar for his day's work.

Back in Cairo, this time staying in a large central hotel, I discovered the need to have a guide always on my side. Whenever I walked out into the street, pairs of beggars would descend menacingly behind demanding baksheesh to go away. Later I was not prepared for the extraordinary, virtually blinding, gold ornaments that dominated key rooms of the Egyptian Museum. By itself, this museum seemed worth the effort of coming to Egypt. Luckily I did not yet feel the dysentery bug already acquired somehow along my travels. Intestinal spasms first hit me in the departure lounge at Cairo Airport and dominated my two-day stopover in Germany to see how Alfred Gierer's RNA infectivity experiments had progressed. By then I was more than ready to be on the Cambridge scene again, where an English-speaking doctor could give me pills to end the feeling of war within my lower parts. Unfortunately, in so crying for help, I got not pills but commitment to what proved a senseless week's stay at the "Pest Hospital" on Mill Road; the possibility of my infecting a Clare oarsman preparing for the late May "bumps" was nct to be risked.

30

Cambridge (England): May–June 1956

I HAD JUST been released from the "Pest House" when Celia Gilbert breathlessly opened the door of my Clare room to declare she had "found her." Not having any idea what she meant, I got up from my prone position to offer Celia tea as well as chocolate biscuits. Too rushed to stop for more than a few minutes, Celia's purpose was to invite me to dinner the next night. A daughter of a rich friend of her journalist father, I. F. Stone, was coming for Sunday-night supper. That her name was Margot Lamont made me realize that Celia had indeed located a rich girl. Her father was Corliss Lamont, a son of the celebrated banker and dominant figure in J. P. Morgan during the 1920s. Well known as the richest American sympathetic to left-wing politics, Corliss's academic bent made him a natural friend of the intellectually inclined Izzie Stone, now a columnist for the left-leaning *New York Post*.

Wally and Celia were living in Green Street, in a flat above the small coffee shop next to The Whim. When next day Celia introduced me to Margot, we started to probe each other's pasts. Right away Margot apologized for a temporary problem with hearing in one ear, which slowed down Celia's and my attempts to interject mild sarcasm in questions about her student life in the U.K. at Birmingham University. She had acquired a boyfriend, and her remarks about him made me feel that she wanted to be back with him as opposed to taking more time exploring the building styles of Cambridge's colleges. After supper was over, she went off to her accommodation for the night, letting Celia and I agree that a monied left-wing home did not necessarily generate an unconventional mind.

Being from the Bullard family was quite a different matter. The straight, dark-haired Belinda, 20-year-old daughter of Teddy, the Cambridge geophysicist, and Margaret, who now wrote novels, was definitely a free spirit. Three years before her mother's *Perch in Paradise* caused a minor scandal; the bed-hopping antics of its main characters were too easily assignable to known Cambridge academics. Of an East Anglian family from Norwich, where everyone drank Bullard Ales, Teddy with his family lived in Clarkson Road, whose largish houses implied occupants with incomes more than academic salaries. Just before I was put in the "Pest House," I went to the Bullard home for a Saturday-night party where Belinda told me she was in her second year reading biochemistry at Girton. There she knew Julia Lewis and had met Peter Pauling. Walking back with me to Clare, she helped me climb over the bike shed and then pulled herself over the top. Up in my room she saw that I was equal to nothing more. I told her that I was weak from a bug giving me gyp tummy.

In a note to me in hospital, Belinda wrote that the almost-summer heat had already made many Girton girls' backs red, the first cuckoo of the spring had been heard, and the little metasequoia in the Girton fellows garden had almost tetragonal symmetry. Soon I brought her around to see Celia, who afterwards pronounced her adorable. I knew then, but did not say, that, unlike Christa, Belinda would not cause constant butterflies to tremor through my stomach. Yet she was endearing and comfortable to talk intelligently with.

A week later, I went to the Blue Boar Hotel for morning coffee with Celia's father, on his way back from Moscow, who had checked in the night before. Izzie, in person, was much less the doctrinaire defender of the Soviet Union that I expected from his newspaper columns. When I kept going back to the rape of Czechoslovakia, he had no hesitation in admitting that his free-thinking ways would not let him prosper in a communist-ruled state. I gathered that his wife, Esther, was much less interested in politics, concentrating on looking after Izzie's needs, both as a thinker and a wage-earner. This trait faithfully appeared to be passed on to Celia, equally happy that not a trace of domestic responsibility kept

Wally from devoting all the time he wanted to theoretical physics. With his Cambridge Ph.D. thesis a done deal, he and Celia would be returning in the late summer to the States to be part once again of the theoretical physics group at Harvard. Izzie and Esther had a summer cottage on Fire Island, not too far away from Cold Spring Harbor where I was to spend much of the summer. So I might have more fun disagreeing with him there.

Back in the Cavendish X-ray group, I found John Kendrew, Francis Crick, Ann Cullis, and Don Caspar all enthusiastic about their recent trip to Madrid for an early April symposium featuring the structures of proteins, nucleic acids, and viruses. Rosalind Franklin and her Birkbeck virus group attended, too, as did Maurice Wilkins with his King's College DNA structure compatriots. Of real surprise to Francis, and even more so to Odile, they found themselves not only tolerating but much

At Madrid Crystallographic Meeting, April 1, 1956. From left: Ann Cullis, Francis Crick, Don Caspar, Aaron Klug, Rosalind Franklin, Odile Crick, and John Kendrew

liking Rosalind. In company that did not make her nervous, she was fun to have around. Peter Pauling, before his problems crashed in on him, had long planned to be there also. With Julia in the easier middle part of her pregnancy, at the last moment he also made the trip. As newly marrieds, they were living south of the Thames, near the Elephant and Castle Station from which he could take the underground into the Royal Institution where he had already started working.

The prospect of solving the structure of polyadenylic acid, or poly(A), was now excitedly bringing me regularly to the Cavendish X-ray room. As a synthetic polyribonucleotide, I thought it might provide vital clues about RNA's structure. On my return to Cambridge from Egypt I heard that some highly purified poly(A) had just been made in Roy Markham's lab, using procedures published by Severo Ochoa's lab in New York. Using it, I made an oriented fiber and put it in front of our rotating anode X-ray beam. Happily its X-ray diffraction pattern had the smell of a helix. An even better pattern later emerged from a fiber made just before Belinda and I strolled to the nearby Arts Theatre to see a weekday performance of Samuel Beckett's new play *Waiting for Godot*. With so little said between its two tramp protagonists, I became bored and restless, eager to get back to the Cavendish basement to see the latest X-ray pattern from this RNA-like molecule containing adenine as the only base. It convincingly told me that poly(A) was also a double helix. Its two chains had their sugar-phosphate backbones on the outside and running in the same direction, each making a complete turn every 31 angstroms. Two hydrogen bonds between pairs of bases held the two chains together with their base-pairing arrangement identical to that earlier found in crystals of adenine hydrochloride. Now that I had poly(A)'s structure right, I could go back to the States with the satisfaction of again doing solid crystallography.

By then I knew that I would have Alfred Tissières with me at Harvard. His research fellowship at King's was ending next year, and he saw advantages to closing down his work on biological oxidation to investigate instead how RNA carries the genetic information for protein synthesis. Several years before he had briefly used the ultracentrifuge in Markham's lab to examine the small RNA-containing cellular particles on which protein synthesis occurs. These particles were the size of the small spherical plant viruses and, like them, might be constructed from large numbers of identical protein subunits. At Harvard, we planned to characterize them in much more detail. That my new lab in the States had to have a primary experimental bent was all too clear. Even Francis was tired of speculation carried on too long before the relevant experimental facts were known. Now Alfred had to decide whether he should take his Bentley across the Atlantic.

Late in May, I began writing my manuscript for a meeting entitled "The Chemical Basis of Heredity" that was to be held in Baltimore at Johns Hopkins University in mid-June. Francis and I were to give separate talks, he on DNA and I on poly(A). Still owning few possessions or clothes, it would take less than a day to pack my belongings for my return to the States. The Pauling–Corey space-filling atoms, which I had used in vain to find the RNA structure but had just now successfully employed in finding the poly(A) helix, I would give to the unit. There was no good scientific reason to take their heavy bulk on to Harvard.

Just before I left for Boston, Belinda Bullard and I slipped without tickets into the half-finished May Ball of Clare, my rented black-tie clothes helping me not to look out of place. This year, the May Ball was held in mid-June, the week before the end of the term. Half-alcoholic by the time we sneaked in, the night was just warm enough to stay through to the early morning hours. With time our dancing became increasingly droopy, and we found a bench to sit on and hold hands before the last dance made us get up. This was my first May Ball and it had not ended the way I once wanted—with Christa. But as the dawn broke, I was not unhappy as Belinda and I made our way towards breakfast.

31

Baltimore, Cold Spring Harbor, and Cambridge (Mass.): June–September 1956

UPON MY ARRIVAL at the 10-storey, red-brick, 1920s Baltimore hotel, I discovered that the McCullum Pratt Meeting organizers had assigned Francis and me to a large, presidential-size suite on the top floor. With its elegant French-styled furniture, we had never before been so well treated, and Francis beamed, pointing out that we were getting the recognition that the double helix deserved. The meeting itself was equally first class. Its organizers had assembled speakers and an audience appropriate for the scientific world's first double-helix-based overview of genetics. George Beadle ("Beets") was there from Caltech to open the meeting, and he gave a very intelligent overview of what genes are, how they might be replicated, and in what way they direct protein synthesis. Long known for his association with the one-gene, one-protein concept, Beets now spoke of genes as segments along DNA molecules that most likely specified RNA templates for ordering amino acids in proteins.

Geo Gamow was missed at the meeting as was his genetic-code theorizing. At this moment, he was having fun in San Diego at the space branch of Convair, doing long-range rocketing calculations that included a trip around the Moon. His long-awaited divorce was finally about to happen with the alimony settlement signed and sealed allowing Rho now to be in Reno. Afterwards he went to the Boulder Summer School, from where he later joyously wrote Alex that he had just been asked to join the University of Colorado's Physics Faculty and was about to go to a "champagne party" that he was giving to celebrate.

That viral RNA carries genetic information became established fact with this symposium. Heinz Fraenkel-Conrat gave the Berkeley results that confirmed the Tübingen data given in the subsequent discussion by Gerhard Schramm. There were, in addition, two potential bombshells. One was the report by Elliot Volkin and Larry Astrachan from Oak Ridge that the unstable RNA made after phage T2 infection had a base composition very similar to that of T2 DNA. Years later, we realized what they were studying was the very RNA made off DNA templates involved in protein synthesis. Their talk pushed the alternative idea that this RNA was a precursor on the way to being transformed enzymatically into DNA. Much more exciting to the audience was Arthur Kornberg's report that his St. Louis lab had possibly observed genuine DNA synthesis in extracts prepared from *E. coli* bacterial cells. The precursor nucleotides involved and the enzymes to pop them together were possibly up for grabs.

The session on nucleic acid chemistry was opened by Erwin Chargaff, who said that too much attention was now being given to nucleic acids at the expense of other worthy cell constituents, such as proteins, polysaccharides, and lipids, further warning that the deeper nucleic acids entrap us, the darker it becomes. In following him, Francis, then wearing his RNA tie, wisely did not rise to Chargaff's caustic bait. He stuck almost schoolmaster-like to DNA's conformational details, emphasizing that for any molecular model to be seriously considered, its creators must build a stereochemically satisfactory proposal. Here he slapped down another jealous biochemist's remark earlier that "real research is done at the bench and not by playing with metal models." Only briefly did Francis allude to possible codes, not having the heart for speculations without facts.

When my turn to speak came, I realized only too well that my poly(A) structure had no real significance for the assembled audience because real RNA itself could not have a similar structure. More exciting was Alex Rich's report that poly(A)–poly(U) double helices form if samples of poly(A) and poly(U) are mixed. Conceivably the resulting double helices are DNA-like, but Alex thought his X-ray patterns favored the

*Francis Crick wearing his RNA tie in
Baltimore, 1956*

two chains running in the same direction like poly(A), not in the oppo-site direction like DNA. In either case, Alex relished Julian Huxley's recent reference to his result as molecular sex—two molecules that embrace each other as soon as they meet.

At the meeting's end, I drove back to Washington with Alex, where my brother-in-law Bob Myers would meet me to take me to his and Betty's home in Falls Church, near where he worked for the govern-ment. Soon they would be returning to the Far East, this time to Indo-nesia, for work in the U.S. Embassy at Jakarta. When their newborn second child, Holly, was able to travel, they would be off. Knowing that Bob's 1954 MG-TF had to be sold, I wrote a check for his asking price of $1500 and drove it several days later to Cold Spring Harbor.

With most students gone, summer was not the time to be in Cam-bridge, Mass., at least in comparison to Cold Spring Harbor where many clever visiting scientists would be in residence. Salva Luria was teaching the Phage Course that was following a new course on Fungal Genetics for which Guido Pontecorvo was the main instructor. When I arrived I

At the wedding of JDW's cousin, Ruth Watson, to John Martin in New Haven,
November 1955: (from left to right) William, James Watson Sr., Margaret Jean
Watson (mother), Betty Myers (sister), and Betty Watson (aunt). The car is the MG
that JDW later bought from his brother-in-law, Bob Myers.

found it already in progress and Ponte anxious during breaks to play
Ping-Pong under the Blackford Hall dining porch. Also back, after several years of absence, was Susie Mayr washing dishes for the summer
courses. Anxiously, I asked about her sister Christa and learned that
over Easter she had travelled deep down into Italy and was expected
back in late August for several weeks of rest before starting her senior
year at Swarthmore. Susie knew that as the summer ended, I would be
on tenterhooks, but she was not now privy to Christa's confidence and
didn't want to predict her arrival mood.

    In Jones Laboratory, I had quickly set up a primitive phage lab, helped by my first graduate student, Bob Risebrough. Born in Canada, Bob had gone to Cornell University as an undergraduate because of an interest in ornithology. Now he wanted to move on, like I had almost a decade earlier, to experiments with phages. I suggested he spend the summer learning how to work with the very small phage $\phi X174$ that might have an equally small DNA molecule. Luria's lab in Urbana still had this little-studied phage in storage, and a sample was quickly sent so that we could soon prepare it in large amounts. Jones Lab, however, proved not the place for phage purification, although Bob isolated several phage mutants for study over the rest of the summer.

A letter I posted back to Belinda Bullard in Cambridge, England, was quickly answered with the news that she had just visited Peter and Julia Pauling at their Clapham home in London. They had moved into larger quarters to give Peter space to be far away from their forthcoming baby. It was now due in mid-August. Belinda thought their new home nice except for two rooms in the middle—in a muddle that Julia, who was always tidy, could not have created. Sprawled across them was a huge electronic set-up that served as Peter's gramophone. Not wanting to spend the summer under her mother's eye, Belinda first eyed Alice Roughton's house in Adams Road. Diverse paying guests stayed there, so she thought it might be an amusing experience for the summer. When she visited, its door was open and she popped in to find a son of the family, Geoffrey, in black trousers, a stiff shirt, and a black bowler hat. His defunct poetry magazine had been called *Oasis,* but now he did statistics of some sort. Most of the rooms had more than four guests, and two old Rolls-Royces were in the garage. Next door lived Alice Roughton's sister with seven children and a bearded husband, who once wrote a book on breadmaking. In the end, Belinda thought it prudent to rent a room at 27 Green Street, across the street from Wally and Celia Gilbert at No. 24.

Further Green Street news came from Celia, who reported that on the July day she wrote they had been vouchsafed 10 minutes of sunlight with the weather freezing all the time. Happy to have gotten my unexpected letter from Long Island, she jumped at my offer to make her my technician when my Harvard lab started up in the fall. Pleased that I so magnanimously wanted to employ a cretin, as she put it, she wondered what rules might exist about their employment permission from a doctor or nearest relative. Here I most certainly had Wally's wholehearted approval for it would mean Celia learning some science and making her, in his eyes, a more worthy mother of Gilberts. Much less important than Wally agreeing was the generous salary that would come out of my NSF grant. Celia had already immersed herself for half an hour in a Belinda-lent chemistry book where she found the words *covalent bonds, ionic bonds, anions,* and *cations,* the latter of which sounded suspiciously to her like a dirty word.

Having read Russian short stories and felt how wretched a sinner she was, Celia was on to Balzac and appreciating how viciously and terribly nearly everyone else behaves. From a BBC radio program on Henry James, she knew his agony in deciding how to address the Army and Navy store in London to order six pounds of Oxford marmalade. For the boat voyage she was holding back reading his *Tragic Muse* to read it on, ugh, the "D" deck. Then when home, she must show off to me her masterful gingerbread.

Back in Cold Spring Harbor, I increasingly looked forward to a visit up from Washington by Alex and Jane Rich. Alex would then let me know whether he still thought the two chains of his poly(A)–poly(U) double helix ran in the same direction. To Jane, I could let out my increasing jitters as Christa's return-home time got closer. They were coming up for the wedding of the daughter of her Ames aunt, whom I had met earlier in London and who lived with her family in a house befitting the daughter of a J. P. Morgan partner. The wedding was to be at St. John's, the beautiful wooden-steepled 1840s church, built by the Jones family, then Long Island's most prominent family. Until Jane arrived, I had hoped to crash the wedding reception, if not the wedding itself, to look over daughters of the lab's wealthier neighbors. Jane, however, said no, her mother would not like me stepping around her family's social code.

But she gave me the cheerful alternative of meeting later that evening Ann McMichael, the pretty small blonde who had so captivated me the summer before on the shore of Lake Geneva. That Ann then had so warmly responded to my bug-sickened form made me wonder whether her marriage was that strong. So I was not surprised to hear that she had separated from her husband and the marriage was soon to end. But I didn't warm to the news that Ann had a new male friend and was anxious especially for me to meet him.

I knew that Dick Feynman was nearby, talking with high-energy theoreticians at the Brookhaven National Lab further east on Long Island, so we asked him to join our party. Several weeks before I had gone there to have supper with him and learned that his marriage to Mary Lou had fallen apart. An art historian, Mary Lou wanted life with a dignified professor, who wore ties and with whom she would look in place in her high heels. The drums and Dick's unrestrained antics at student-dominated parties increasingly rubbed her up the wrong way, and earlier this year he could take no more of their constant arguing. On the phone I told him that a pretty, now-free girl I had met in Switzerland would be there for him to gaze upon.

At the Chinese restaurant in which we gathered about a large, round table, Dick could only be his volatile, show-offish self. It was virtually impossible for him to enjoy himself without being the life of the party. When the long supper was ending, Dick took me aside and asked what I saw in Ann. Gone, at least temporarily, was the youthful thirst for life that let Ann share hands with me in an apple orchard beneath the Alps.

August was in its last week when, on a Tuesday evening, Susie Mayr told me that terrible news had just come in a letter to Ernst and Gretel. I instantly feared that Christa had fallen in love with a German student and would not soon be coming home. But the news was much, much worse. Christa's letter revealed that she was pregnant and that the father was an engineering student she had met in Munich. The baby was in no way coming by design. Nonetheless, she wanted to have the child and marry the father in Germany after a month's visit home to see her parents and Susie. Most importantly, she was not doubting her decision and did not want to come home to give her parents the chance to try to change her mind.

Susie was virtually in tears as she talked, and I became too choked up to reply coherently. Learning that Ernst was soon leaving their New Hampshire farm to come down to Storrs in Connecticut for the annual meeting of the Genetics Society, I arranged to meet him two days hence. There I found him fatalistic, telling me how strong-willed Christa could be and revealing that her letters over the past eight months had much worried him. Particularly bad was her going alone to Italy over Easter, with so little concern about falling into danger, personal or otherwise. In his opinion, she was much too immature to have a baby by a youth that she had known for such a short time. But he did not know how to convince her of this. After she was home, now in only a few days, he and Gretel would do the best they could to keep Christa from ruining all her chances for a meaningful future.

Continuing on to Harvard later that night, I went to Paul and Helga Doty's new home on Kirkland Place the next morning to tell my bad news over coffee. Paul warmly tried to reassure me saying that, with my new life at Harvard about to start, I was bound to find an even more suitable girl from Radcliffe, the women's college near Harvard. Helga then, to my immediate pain, told me that with time I would realize my news was not so bad. For reasons she couldn't define in words, she had long felt that it would never work out well for me with Christa. Far better that it was now over definitely than to go through more years of emotional turmoil.

I had to see Christa before she left and did so one evening when she had been back for a week. If she had weakened in her determination to have the baby, by then it would have happened. Ernst and Gretel left us alone and we had the living room to ourselves. Both of us tried to smile and talk like good friends catching up on each other's last six months. The boy she was marrying was already a trained engineer who wanted to work on nuclear reactors and she felt confident for his future. Most importantly, he was not tense and almost instinctively knew how to keep her happy. That she knew him so briefly did not matter. He was good for her. There was nothing I could say in reply: Christa had made her choice. To argue that she was acting stupidly would have done no good, and my visit would have ended on an angry note. After wishing her well and not being sure but shaking hands, I was off into the street. I got into my MG and slowly drove to Divinity Avenue, where I parked outside the Biological Laboratories that were to prove my effective home for many years to come.

Not capable of being alone then in my new office, I walked towards Harvard Yard, where just a trace of early-fall evening light remained.

Slowly I calmed down, the outlines of the giant elms above me giving the feeling that those who chose to walk beneath them go on to rightful destinies. The fall katydid sounds were everywhere, and I was virtually alone as I walked back and forth along the yard's diagonal paths not wanting to go further out into the noise and light of Harvard Square.

Epilogue

October 1956–March 1968

THE PERSONAL AGONIES that for many of us dimmed the glamour of the double helix's first glory days did not last much longer. For the most part, they were replaced by much happier emotional commitments. Likewise, the intellectual disappointments that accompanied our attempts to jump beyond the double helix before experimental facts had firmly gripped our feet soon ceased. Just seven years were to intervene between the 1953 finding of the double helix and the 1960 understanding of the basic ways that RNA is involved in protein synthesis. With messenger RNA and transfer RNA at our disposal, the genetic code took only six more years to crack and complete.

Mariette Robertson left England, in June 1956, for Paris, where she spent the following year at the Sorbonne. But she returned permanently to her Caltech roots, marrying the Harvard- and Oxford-trained historian of India and the Opium Wars, Peter Fay, in 1958. Later, the couple and their four children spent two years in India. Meanwhile, Linda Pauling married geophysicist Barclay Kamb, recently appointed to the Caltech faculty of which later he served for several years as its Provost. In 1964, after Linus left Caltech, the Kambs and their four sons moved into the foothills home where Linda was raised.

Linus's departure from Caltech followed the awarding to him, in late 1963, of the Nobel Peace Prize in recognition of his efforts to prevent the use of nuclear weapons. His widely acclaimed second Nobel award, sadly, was ungraciously received by Caltech's trustees, who perceived him as aiding communist causes. By then Linus had become increasingly frustrated by Caltech's cutbacks of his research space, and he accepted a position at the Center for Democratic Institutions in Santa Barbara directed by Robert Hutchins.

Av Mitchison was also married soon, to Lorna Martin, a student at Edinburgh University. Her parents lived on Skye, to which I went in July 1957 to be Av's best man at the ceremony in Portree. I took my father, much at loss after my mother's sudden death in May of that year, to the wedding. Mother's heart, long faulty from a childhood streptococcal infection, could go on no longer. My Harvard lab had moved by then into high gear through the arrival the past March of Alfred Tissières and his Bentley—all too soon to be sold to a Harvard Law professor. Celia

Av and Lorna Mitchison after their wedding on Skye in August 1957;
JDW was best man. Nou Mitchison is second left.

With Alfred Tissières in front of the Biology Laboratories at Harvard

Gilbert, by then, had retired from her job with me and took comfort that she no longer had to think in terms of factors of 10. Instead, she was happily anticipating the impending birth of her and Wally's first child.

Then living near Harvard was Margot Schutt, whose Henry James manners so fascinated me on the boat that brought me back from England as the summer of 1953 ended. She was working for the book publishers Ginn & Company. In mid-April I drove her to Woods Hole to let Marta Szent-Györgyi see me with a girl who knew the consequences of vulgarity. The meeting between these two ladies—one young, one middle-aged—did not go well, in part because Margot had not thawed from the wind rushing through my transparent MG. Whether my unexpected re-emergence into Margot's life had anything to do with her decision a month later to take a job in London, I was not given reason to know.

Over the 1957–58 academic year, Alfred and I found that the *E. coli* ribosomes were not structurally like spherical RNA viruses but had a bipartite structure with one subunit twice the size of the other. In the presence of sufficient magnesium ions, the two subunits stick together but fall apart at lower Mg^{2+} ion levels. Subsequently, we were puzzled by observations that both subunits contain single RNA molecules, that of the larger subunit being twice the size of the one within the smaller subunit. We had expected to find heterogeneously sized ribosomal RNA molecules, reflecting the greatly different lengths of the polypeptide chains they reputedly coded for.

In July 1958, Alfred married the spirited Virginia Wachob, who had been living in Paris but was from Denver. I was a best man again. Colorado also did well for Geo Gamow, who after arriving in Boulder in 1956 wanted to hold a molecular genetics meeting in August the following year. In December 1956, he wrote with more details of the conference and mentioned he was finishing off his latest book, *Matter, Earth and the* *Stars.* Two years later he made a second marriage to the similarly aged Barbara Perkins. Her past life had been in publishing and over the next 10 years she assisted Geo in the publication of several more books. In 1968, aged only 64 years old, he died prematurely from alcohol-induced liver failure.

Just after I left England to go to Harvard in the summer of 1956, Sydney Brenner left South Africa to join Francis Crick in Cambridge, adding extraordinary genetic zip to the Cavendish site. During the next few years, experiments both from my lab and Sydney and Francis's, where Leslie Orgel's wife Alice also contributed significantly, began to show how mutagenic agents act at the DNA level. At Harvard, key work was done by Ernst Freeze, who had trained as a theoretical physicist in Germany under the famed Werner Heisenberg and had afterwards moved into biology through a year with Seymour Benzer at Purdue. There, Ernst showed that the sites of 5-bromouracil-induced mutations differ from those where spontaneous mutations occur. After coming to Harvard, he went on to demonstrate that two quite different types of mutation exist. Those caused by base analogs he interpreted correctly as "transitions" (replacements of one pyrimidine by another pyrimidine or one purine by the other purine). The other class of mutation, induced by DNA-binding dyes, like proflavin, he called "transversions," thinking them substitutions of purines for pyrimidines, and vice versa. Here time would prove him wrong.

Early in June 1959, Francis, Sydney, and Ernst came together for a meeting at Brookhaven National Lab. Francis and Odile were temporarily living in Cambridge, Mass., staying with the Riches, while Francis was a visiting professor in Harvard's Chemistry Department. The Riches had by then been ensconced in a Linnaean Street house for some six months after Alex's move from the National Institutes of Health in Bethesda to MIT. Francis radiated enthusiasm about the now-universal acceptance of his "Adaptor Hypothesis." No one, including himself for a time, thought it on the right track until the fall of 1956. Then, soon after my arrival at Harvard, Paul Zamecnik at the Massachusetts General Hospital told me about his studies with Mahlon Hoagland showing that amino acids on the way to being incorporated into proteins are first covalently bound to RNA molecules. Mystified by what their result meant, I suggested to them that they had discovered the "RNA adaptors" that Francis had postulated in his 1955 note to the RNA Tie Club.

In the spring of 1959, Aaron Klug was having difficulty in being admitted to the States. Rosalind Franklin had died the year before at the young age of 37 from the ovarian cancer that had been diagnosed after an operation 18 months previously. After her second such operation, she recuperated in the home of Francis and Odile. She felt more comfortable with the Cricks than with her parents, with whom she too easily found herself at odds. Aaron, now in charge of Rosalind's former lab, wanted to share ideas again with Don Caspar, by then at the Children's Hospital attached to Harvard Medical School. But Aaron's supposed political past in South Africa had made him suspect by our State Department officialdom. I wrote on his behalf to the U.S. Embassy in London, hoping to reverse their paranoia. To the relief of those of us embarrassed by our country, he soon got his visa.

Later in 1959, in August, a meeting in Copenhagen brought together 30 of the key figures trying to unravel how DNA provided the genetic information for protein synthesis. In his talk, Jacques Monod emphasized the puzzling results that he, François Jacob, and Art Pardee had just obtained at the Institut Pasteur in Paris. They had found that the synthesis of an induced enzyme takes off maximally, within minutes of its coding gene's entry into a cell. If new ribosomes have to be synthesized, he said, several subsequent cell divisions would have to occur before synthesis at the maximum rate. Then Jacques wondered whether induced enzymes represent proteins that are directly made on DNA templates. To me, this was a horrid possibility.

Much easier to interpret were the elegant experiments done the year before at Caltech by Matt Meselson and Frank Stahl, which showed unequivocally that the two strands of the double helix separate during DNA replication. As soon as his Ph.D. thesis with Linus Pauling was complete in 1956, Matt focused with Frank on how heavy isotopes might be used to distinguish parental from progeny strands. This they achieved by separating parental *E. coli* DNA labeled with the heavy isotopes ^{15}N and ^{14}C from progeny DNA labeled with the lighter ^{14}N and ^{12}C isotopes in density gradients of the heavy salt cesium chloride. Experiments done soon afterwards by Julius Marmur in Paul Doty's Harvard lab were almost as exciting. They showed that separated single strands, made by exposing double helices to high denaturing temperatures, can subsequently come together under lower annealing temperatures to form hydrogen-bonded double helices.

The new technique of sucrose gradient centrifugation had by then begun to dominate ribosome studies in our lab. Developed in Washington by the Carnegie Institute biophysics group, sucrose gradients allow centrifuged extracts to be sampled for radioactively tagged cellular components. By the fall of 1959, at Illinois University, Masayasu Nomura and Ben Hall, Paul Doty's former student, were using them to look at the "DNA-like" RNA synthesized after phage T2 infection. Because of Masayasu's impending move to Seymour Benzer's lab, they had little time for experiments and thought that T2 RNA became part of a smaller ribosomal subunit. When I read the resulting manuscript, I was not so convinced and had my first graduate student, Bob Risebrough, do further experiments early in 1960. Within two months, he showed that ribosomes are not templates but are instead molecular factories to which T2 RNA RNA (later to be called messenger RNA, or mRNA, by Jacob and Monod) binds as a template and directs the ordering of amino acids during the synthesis of polypeptide chains. We now saw why our previous three years' work on ribosomes had yielded so many unwanted results and we began a search for mRNA in uninfected *E. coli* cells.

Such experiments began in late May 1960 after François Gros arrived at Harvard from the Institut Pasteur. Soon he and I were joined by Wally Gilbert, now an assistant professor of physics. Learning of mRNA over a Celia-cooked supper, Wally, then fearing that his "dispersion theory" forays in physics were going nowhere, thought he would have much more fun trying to find the templates that order amino acids within bacterial proteins. By the time our first such *E. coli* experiments succeeded, we knew the "ribosomes as factories not templates" concept had been independently demonstrated at Caltech. Earlier, a mid-spring visit to Cambridge, England, by François Jacob brought him to Sydney Brenner's rooms in King's. There they, with Francis Crick, batted about the idea that short-lived RNA molecules, like those synthesized after T2 infection, might be templates for the induced enzymes then under study at the Institut Pasteur. Excited by their possibly important brainstorm, Sydney and François went soon afterwards to Matt Meselson's Caltech lab. Through his cesium chloride sedimentation tricks, they also found that T2 RNA binds to ribosomes synthesized before viral infection.

The fall of 1961 was for me dominated by the presidential race between John Kennedy and Richard Nixon. I joined Alfred and Virginia around their TV set to see the presidential debates. After Kennedy's victory, McGeorge Bundy went with him to Washington as the White House National Security Advisor. Paul Doty and I breathed a sigh of relief that Bundy had not left Caltech earlier, because only with him as Dean would Matt Meselson have been offered and have accepted a position in Harvard's Biology Department. Matt was now married to a girl he met at the Aspen Music Festival and had arrived at Harvard just before Dick Feynman came through to give a Biology Seminar in early February 1961.

I had previously learned from Matt that Dick was about to get married and that he had become a part-time biologist doing phage experiments. So I wrote Dick asking him to come to Harvard to give a talk about his work. He quickly accepted, writing me during his honeymoon in San Francisco and ending with his RNA Tie Club code name "Glycine." He and his new wife, Gweneth, had met first on a beach near CERN, the big European physics lab outside Geneva. She was temporarily away from her au-pair responsibilities when Dick first spotted her. He urged her first, unsuccessfully, by mail to come to Pasadena as his au pair. Now he had the wife that would be his for the remainder of his life.

Harvard was covered with a foot of new snow the day Dick arrived in early January, causing me to rush to the Riches' to borrow the snow boots that made Harvard so different from Caltech. Later, in his lecture on February 6, Dick talked about new experiments with phage T4 *r2* mutants induced by proflavin, an acridine dye. To his surprise, he found that the revertants back to wild types themselves were not mutations at the same site but suppressor mutations at nearby sites in the same gene. Interestingly Dick had noticed that the *r2* suppressor mutant, when alone, had the same phenotype as the mutants they suppressed. Present together, they somehow neutralized each others' effect. Later that spring, I arranged for Geo Gamow to lecture on the "Origin of the Universe." The large audience on April 26 almost filled the Burr Hall lecture room with students attracted by his "Mr. Tompkins" popularization of physics and astronomy.

Six weeks later, the June 1961 Cold Spring Harbor Symposium was dominated by messenger RNA's recent discovery. The "ribosomes as factories" papers from Harvard–Pasteur and Caltech–Cambridge had just been published in *Nature,* and soon to appear would be the long review paper by Jacob and Monod in the *Journal of Molecular Biology.* Brand new, and particularly important, was the use of Julius Marmur and Paul Doty's DNA annealing procedures by Sol Spiegelman's lab in Illinois to make hybrid double helices containing DNA chains hydrogen-bonded to their mRNA molecule products.

In late July 1961, Peter Pauling stayed with me in my Appian Way flat. He was in good form, his unwanted move five years before to the Royal Institution being a blessing in disguise. There, he got out of protein crystallography and did an inspired thesis on inorganic molecules that led him to a lectureship in Chemistry at University College London. He and Julia had recently had a new daughter, Sarah, to complement their son, Thomas, and were living up from the Notting Hill Gate tube station in Lansdowne Crescent. Stricken with tonsillitis upon his arrival, he went for help to three Radcliffe College girls ensconced for the summer in the small home that my father, now living near me, had vacated for the summer. The sweet Emily had no idea what she was doing when she offered to help nurse Peter through his recovery. Not wanting to know how this care was given, I felt relief when Peter flew on to Denver for the annual meeting of the American Crystallographic Association.

Peter's research was then supported by the Office of Naval Research

Fairmont

HOTEL · ATOP NOB HILL · SAN FRANCISCO

Dear Jim,

Matt's right — I've gotten married & am now on my honeymoon.

It would be so much fun to see you again that I can't resist your invitation. I am also working on RNA — half time, and I would like to try to do experiments with you & the bevy of beautiful lab girls.

But do I have to give a lecture to the Biology Dept? What do I know that they don't know about biology? — nothing. I can't think of anything I could say.

Do you want me anyway?

OK, then — my wife & I will come to Cambridge on or about Jan 1 to work for 3 or 4 weeks. A deal?

Thank you very much.

Glycine

[alias, R. P. Feynman.]

Peter Pauling on a visit to JDW's office, c. 1966

(ONR), whose travel orders often let him get lifts on military airplanes. Going on to Seattle for more lectures, he used his travel orders to get on a flight to San Diego, but not being on the manifest, upon his arrival he was fingerprinted and questioned by security personnel as to his right to be on military aircraft. Finally convinced that he was not defrauding the Federal Government, they let him hop on a bus to Los Angeles and his parents' home above Pasadena. Later, the sum of $135.51 flew him to Pittsburgh for more crystallography and a further $14.41 got him to Washington, where at ONR he got further travel orders to London, soon exchanged for a TWA ticket to London. Peter wrote me that perhaps he is an army man after all.

At the beginning of August 1961, Wally Gilbert and I stopped off in Helsinki and then Leningrad on our way to the International Congress of Biochemistry being held in Moscow. In the midst of the congress, there was a large parade through Red Square celebrating the triumph of the second Russian cosmonaut, Titov, who had just circled Earth. Luckily I was able to watch it through a window of the adjacent Hotel National. While I and most congress participants were condemned to rooms in the Hotel Ukraine across the Moscow River, Robert Maxwell, whose Pergamon Press was publishing the proceedings, occupied a National Hotel suite once occupied by Lenin. Paul Doty, who knew Maxwell through the Harvard chemist Bob Woodword, thought there would be no problem in my viewing the festivities from Maxwell's rooms. Francis Crick, however, was not sure that he also would be welcome until the heavyset swarthy publisher told him that any friend of Jim Watson was Robert Maxwell's friend too.

No one from abroad expected that this congress would provide anything but a chance to see communism in action. Soon, however, I heard rumors that there might be an unexpected bombshell talk by Marshall Nirenberg from the National Institutes of Health. It was not one of the main presentations, and only a few individuals, including Alfred Tissières and Wally Gilbert, were drawn to its presentation through the title "The dependence of cell-free protein synthesis in *E. coli* upon naturally occurring and synthetic template RNA." Using Alfred's improved recipe for cell-free protein synthesis, Nirenberg and his German colleague Heinrich Matthai had found over the past several months that addition of poly(U) promoted the synthesis of polypeptides made up exclusively of the amino acid phenylalanine.

When François Jacob heard about the experiments from me over breakfast the next morning, he thought I was perpetuating a practical joke. But Nirenberg and Matthai had done their experiments well, and Francis hastily arranged a big lecture on the last congress day that let Nirenberg convince as well as stun most in the audience. From that moment on, it seemed likely that the genetic code would soon be completely cracked through observing the polypeptides made in cell-free systems programmed with appropriate synthetic polyribonucleotides.

Wally and I flew east, not west, from Moscow. He was on his way to Pakistan, where his economist father worked for the Ford Foundation, while I was on a round-the-world ticket that would let me go to the great temples at Angkor Wat with my sister Betty, whose husband was then attached to the U.S. Embassy in Cambodia. En route I wanted an Afghanistan experience and in Kabul learned that fresh melons were not always what they seemed, often having weight added by immersion in the water running under the streets. To stay healthy, I temporarily joined the International Club, where I supped with two Americans who let on they were economists at the Kabul Embassy. Sensing that they might be otherwise, I asked them whether they knew my brother-in-law, Bob Myers? Their negative answer I would have forgotten had I not later bumped into a friend of Betty's on a ferry in Hong Kong. He told me that he was just with colleagues who had met me in Kabul.

After his return from Moscow, Wally's time away from teaching physics became exclusively directed to messenger RNA. He got poly(U) samples from Paul Doty and showed that several ribosomes can function on single mRNA molecules, thereby explaining why there was so little mRNA in cells. At the same time, Alex Rich observed aggregates of ribosomes (polysomes) in hemoglobin with four to five ribosomes simultaneously translating the relatively short messages that coded for the small globin chains with 155-amino acids.

Meanwhile, Francis Crick and Sydney Brenner were completing genetic experiments in Cambridge to prove their hypothesis of the year before that acridine dyes, like proflavin, cause mutations by inserting or deleting base pairs. As 1961 progressed, and not knowing of Dick Feynman's results, they independently found that suppressors of $r2$ mutations when present by themselves are also $r2$ mutants. Whereas Dick thought $r2$ suppression reflected pairs of changed amino acids neutralizing each other, Francis and Sydney believed their results arose from the genetic code being read in groups of bases, probably three, starting from a defined spot. The addition of a base would alter the reading frames beyond, as would the deletion of a base. If they were correct, suppression of addition (or deletion) mutants would lead not to two separate amino acid replacements but to a stretch of changed amino acids between sites of insertion and deletion. Proof that they were right came from their finding that all their proflavin-induced suppressor mutations fell into two classes (+ or −) with ++ or −− combinations never suppressing. And by finding that groups of three nearby +(−) mutations frequently lead to normal phenotypes, they showed that amino acids are specified by groups of three bases (a codon). Sensing the great importance of this result, the editor of *Nature* saw that their resulting, quickly written paper appeared in the last issue of 1961.

Increasingly by then I was preoccupied by the White House Presidential Scientific Advisory Committee (PSAC). Formed during Dwight Eisenhower's last years as a response to the launch in 1957 of the first Soviet Sputnik, it was initially chaired by the Russian-born George Kistrakowsky, long a Harvard professor, whose labs were below Paul Doty's. George, who was then still on the committee as was also Paul, made my day soon after I returned from my world trip by asking if I would help PSAC in its oversight of our biological warfare effort. No obvious red skeletons popped up in my past and by December I had a White House pass letting me into the Executive Office where PSAC's office was now headed by MIT's sagacious electronics expert, Jerry Wiesner. Two floors above and down the corridor were offices for junior members of McGeorge Bundy's National Security Council. In one sat Diana DeVegh, whose job seemed to owe less to her talents as a budding Arabist than to her dinners with the then junior senator from Massachusetts when she was at Radcliffe.

Diana was a senior (in her final year at college) when we first met at a party given by the British-Israeli chemist, David Samuels. A postdoc in Harvard's Chemistry Department, he would eventually assume the title first held by his illustrious grandfather Viscount Samuels. To my surprise he told me that he and Rosalind Franklin had been cousins and that he had admired her attempts to live by her brains alone and not through her privileged family connections I had not known about before.

For much of spring 1962 I was back on sabbatical leave in old Cambridge where I lived in Churchill College as a visiting fellow. The college had recently been formed with Sir Winston's blessing to provide more science at Cambridge and Francis became one of its founding fellows. But when they decided to build a chapel, using funds specifically designated for this purpose, Francis very publicly resigned, saying he saw "no reason to perpetuate mistakes from the past." Later he gave 75 guineas (£100) to the Cambridge Humanists to sponsor a prize essay on "What shall be done with the college chapels?" The winning essay saw several different futures for them, including swimming pools.

Francis and Sydney had just recently moved their labs away from the Cavendish to the newly constructed Medical Research Council (MRC) Laboratory for Molecular Biology at the new Addenbrooke's Hospital site. The Queen opened the new building and on her tour around she received brief explanations of the three-dimensional models of hemoglobin and myoglobin whose structures Max Perutz and John Kendrew had recently triumphantly solved. Initially the double helix was to be presented by Francis, but he balked, saying that the more appropriate person to open the lab was her husband, Prince Philip. So I took on his chore and when asked by the Queen what she was looking at, I took two minutes to summarize DNA and horses.

I returned to the States in time for the annual Cold Spring Harbor Symposium in June. Virtually everyone of importance in the animal-virus world was there. The meeting started with a clever structural paper by Don Caspar and Aaron Klug that intellectually extended the ideas on polygonally organized viruses that Francis and I put forward six years before. Tumor viruses were coming into prominence with Renato Dulbecco and Michael Stoker beginning to explore how the newly found polyoma virus makes cells cancerous. Also present was André Lwoff, whose talk on polio research had its origins in his 1955 visit to Dulbecco's Caltech lab. While standing in the long food line, André, then 60, could not help but notice the checkout girl, the cook's daughter Amy, whose undeniable beauty was so at variance with that of her mother. Towards the meeting's end, he paid her the compliment of saying he wished he was 20 years younger. In reply, Amy said that 30 years would be better.

During the summer of 1962 Alfred Tissières was back at Harvard from Europe where he had just become a professor in Geneva. Virginia went on to Denver to be with her mother for the birth of their first child. Embellishing the lab then was urchin-like Radcliffe student Pat Collinge. I had been told she needed a summer salary and eagerly found her a lab job. Pat's blue eyes proved a magnet that kept me not too far from the lab that summer. Near August's end, I persuaded her to drive with me to Woods Hole before she joined her boyfriend, then out on

Cape Cod. At Woods Hole, in Albert Szent-Györgyi's house, Pat typed out the first page of the book that I initially called "Honest Jim," and whose first chapter began with the words "I have never seen Francis in a modest mood."

In early October, the news came out of Stockholm that Francis, Maurice Wilkins, and I had received the 1962 Nobel Prize for Physiology or Medicine. That Maurice was included pleased both Francis and me, but we had to wonder how the prize would have been divided if Rosalind Franklin had not died so tragically young. Then, as today, the Nobel's rules preclude dividing any given prize among more than three individuals. My sister Betty came with me to the almost weeklong ceremonies, as did my father. We stayed in the Grand Hotel that looks out over water to the Royal Palace. Also there were John Kendrew and Max Perutz, who had been chosen as that year's winners of the Nobel Prize for Chemistry. Geo Gamow's Russian friend, Lev Landau, the 1962 winner of the Physics prize, sadly could not come—not because of communist fears that he would bolt, but a tragic recent road accident that had left him badly impaired mentally and physically. I would never have the opportunity of seeing why Gamow thought our personalities so similar.

Over Christmas 1962 I was at Verbier, the Swiss ski resort that Alfred Tissière's brother helped finance and from there flew to Scotland to join the Mitchisons at Carradale for New Year's Eve. Later, in Geneva, I met John Kendrew and we went out together to CERN (then called Conseil Européen pour la Recherche Nucléaire, hence the acronym, but now the European Laboratory for Particle Physics). Leo Szilard was temporarily there, having precipitously fled Washington, D.C., with his wife Trudy, fearing a nuclear catastrophe at the time of the Cuban Missile Crisis. Leo was keen to get started in Europe a Cold Spring Harbor–like lab that could hold courses and meetings and for important research. He wanted Vicky Weisskopf, in Geneva on leave from MIT, to head CERN and to tell us what problems might come up in founding and then funding a CERN-like organization for molecular biology.

At Geneva, my return flight was delayed due to a snowstorm over southern England, and I found myself next to Janet Stewart, the statuesque Girton student whom I originally met through Linda Pauling. For several years previously I had seen much of her when she and Gidon

Gottlieb shared a flat near Harvard Square on Boylston Street, while he did further studies at Harvard Law School. The year before, though, they had gone their separate ways, neither able to keep each other in the style of life that their intellects needed. With Janet on the plane was a youthful-looking Etonian barrister, with whom she had been on a skiing holiday and whom she later married.

The June 1963 Cold Spring Harbor Symposium saw the big labs of Marshall Nirenberg and Severo Ochoa racing to crack the genetic code through varying the base compositions of their synthetic RNA templates. Using this route, roughly half of the potential 64 codons (AAU, AAC, AAA, AAG, etc.) now had amino acids assigned to them. At the meeting, Wally Gilbert from our lab talked about polyribosomes and how growing polypeptide chains are held into ribosomes by their carboxyl terminal transfer RNA molecules. And Leo Szilard again pigeonholed symposium attendees for facts he later wanted to put more cleverly together. The occasion provided the biggest gathering yet of the RNA Tie Club with five of its members present, including Geo Gamow and his ever-present whisky glass.

That fall, my sister Betty was living again in Washington, her husband having been posted back from Cambodia. When briefly down for a PSAC meeting, I met at their home John Richardson, whose removal the week before as the CIA Station Chief in Saigon was widely taken to mean that the USA no longer backed the South Vietnamese Government. Two weeks later, again on PSAC business in the Executive Office Building, my meeting was interrupted by the news that President Kennedy had been shot in Dallas. Too soon afterwards we were told he was dead. That night I had dinner with the Szilards. By evening, Leo had already removed Jack Kennedy from his mind and was worrying how to influence Lyndon Johnson's future decision about nuclear weapons. I had no wish to watch the funeral cortège pass down Pennsylvania Avenue and was back at Harvard by the time the funeral mass had started. I knew that my return visits to Washington would never be the same.

Bad news came again soon when my father, then only 63, suffered a stroke and it was unclear whether he would ever walk again. But over the next several months he became able to move with a cane and continued living next to me in Cambridge until his death, seven years later, from smoking two packs of Camels a day for 40 years.

Meanwhile, my Nobel monies had gone to make a $17,000 down payment on an early-nineteenth-century wooden house within walking distance of Harvard Square. Paying off the mortgage early was now within my means due to the widespread acceptance of my new book, *The Molecular Biology of the Gene,* published in time for the previous fall's college classes. It was the first introductory textbook aimed at college students to emerge from our DNA revolution. In its first year, royalties effectively equaled my salary as a Harvard professor. So, even more, was I in want of a wife.

Whenever I could go down to New York I would have dinner with Gidon Gottlieb, now practicing law in the city and living with his new wife Antoinette, who originated from Geneva, in a penthouse on Sheridan Square in Greenwich Village. Early in 1965 I mentioned to them that I had just seen in Boston Salvador Dalí's massive new painting *Galactosidal Nucleic Acid—Homage to Crick and Watson.* Jokingly we mused whether Dalí might be the appropriate artist to illustrate "Honest Jim." We knew that Dalí and his wife Gala lived during the winter in the St. Regis Hotel in New York, and the three of us taxied there in the hope of meeting him. From the hotel lobby I sent up a hastily scrawled message "The second brightest person in the world wishes to meet the brightest," and signed it Jim Watson. Within minutes he was in the lobby and, in French (which the Gottliebs understood), asked me to lunch several days later at his hotel's King Cole Restaurant.

Initially I felt awkward over our small table as Gala tried to translate to me Dalí's interest in holograms. Then, without warning, a super-pretty young girl with long blond hair came to our table and told me how pleased she was that Dalí was letting her meet the discoverer of DNA. She was an actress on the popular TV program *Peyton Place,* but as I did not then own a TV set I could not connect her incredibly fetching face and voice to her name. Then all too soon, unfortunately, she left to catch a plane for Los Angeles. Believing her to be at most 15 years old, I felt depressed that I, now 37, was unlikely ever to get to know her. Only two years later did I realize who the blond young angel was. After reading a magazine article about Dalí, I saw him at Knoedler's 57th Street Gallery at an exhibition of his recent paintings. I told him I recognized her, and he smiled as he said "Oh, Mia."

From mid-October 1965 through into the New Year, I was across the Atlantic in old Cambridge to learn more facts from Francis so that I could complete "Honest Jim." Only the final chapter remained unfinished when I was briefly back in Harvard early in January. In only two days I had polished it off, ending with the words "I was 25 and too old to be unusual." Then I flew off for a six-week lecture tour sponsored by the Ford Foundation of East African universities. Afterwards I went on to Geneva for several months in Alfred Tissières's lab, then largely devoted to ribosome studies. There Alfred told me that working for him was an attractive Iranian girl called Nasrine Chahidzadeh, who wanted to go to the States. So I took her out to dinner where she told me about her chemistry studies in Zurich. Then she wanted to know more about the Kennedys, having just met Teddy at a large lakeside house in Geneva. Our dinner had not ended when I let Nasrine know there was a job for her at Harvard if she wanted it.

Bringing me back to the States in early June 1966 was the Cold Spring Harbor Symposium on the Genetic Code. By then, all the codons had been solidly established, helped much by Gobind Khorana's combination of chemical and enzymatic tricks for making synthetic RNA of known repeating sequences. This symposium was very much one for Francis Crick to dominate—rather as I held forth at the 1953 gathering and Jacques Monod and Sydney Brenner did in 1961. The tone was set by Crick's opening address entitled "Yesterday, Today and Tomorrow." At the meeting's end there was a mood of great triumph at the cocktail party on the Blackford lawn, which coincided with Francis's fiftieth birthday. Knowing it was a more than noteworthy occasion, I had driven earlier with Paul Doty's student Bob Thach to Entertainments Unlimited in Levittown. There, from a book of pictures, we chose "Fifi" as appropriate for a coming-out-of-a-cake-like act under the Blackford porch. Luigi Gorini, in the know, had his camera on hand to record Francis's laughing reaction to his birthday present.

Increasingly the potential gaiety of any such gathering was diminished by the thickening cloud of the escalating American involvement in Vietnam. We felt that Harvard itself was in trouble. To try and explain what our White House policy was, in June 1965 McGeorge Bundy came to Harvard and spoke to a packed Memorial Hall student audience. Sitting next to Paul Doty, I felt at first dissatisfied with Bundy's answers and, later, sad, not wanting to believe that my former Harvard protector had become a spokesman for a war that could never be won. To my relief, Betty's husband was no longer associated with our Southeast Asia doings. Long wanting to have a try at journalism he had just become the publisher of the new *Washington Magazine* that he founded with his University of Chicago roommate, Lauchlin Phillips.

Early in September 1966, I went to a NATO-sponsored meeting on the Greek Island of Spetsai, a two-hour boat ride from Piraeus, the port of Athens. Francis, with Odile and his two daughters, boated over from Italy, where he was keeping his newly bought motor cruiser. Largely in control, Francis nonetheless had several anxious moments docking in

At the Cold Spring Harbor Symposium, June 1963: (from left to right) Francis Crick, Alex Rich, George Gamow, JDW, and Melvin Calvin

the harbor of the large white resort hotel. Earlier, using Nobel Prize monies, he was temporarily part owner with the wealthy molecular biologist, Gianpiero DiMayorca, of a large sailboat in Naples. Seeing him once at its helm, Jacques Monod later commented this was the only occasion where he'd ever seen Francis in a modest mood.

Harvard University Press then wanted to publish "Honest Jim," its editor, Tom Wilson, liking it greatly when he read it the day after I finished the last chapter. In the fall of 1966, he sent copies of the manuscript and release forms to the principals mentioned in the story. Although Francis and Maurice Wilkins were annoyed by what I'd written, and had so informed the Press's President Pusey, we hoped they would later accept the public's need to know how the double helix was found. We felt more confident about going ahead when we got the Foreword from Sir Lawrence Bragg. The previous March, when I showed Bragg the manuscript in London, I had surprised him when I asked whether he would consider writing the Foreword. The year before that March meeting, not knowing I was already at the task, Bragg wrote me that I should tell my side of the story. Notwithstanding Bragg's very positive Foreword, Harvard University Press was told in May 1967 by President Pusey to reject the manuscript. By then, Tom Wilson did not mind because he had decided for personal reasons to move to New York as a Senior Editor at Atheneum Publishers. In February 1968, they published the book as *The Double Helix*.

Nasrine's Persian beauty did not long grace Harvard's Biological Laboratories. Given her long weekend absences, I was not too surprised when in early March 1967 she revealed she was soon to be married to a Swiss chemist, then working in New York in his family's business office. But I did not expect her then to ask me to give her away in the church ceremony already scheduled in May at Harvard Memorial Church. Apparently her father, for reasons that she did not explain, would then be detained in Tehran. Later, I learned that he had been the lawyer of the former Premier Mossadeq, long a bitter enemy of the Shah.

Not long before the mid-May wedding, on my way to Turkey for lectures again under Ford Foundation patronage, I stopped off in Geneva and popped into the antique shop of the woman in whose home Nasrine had lived before coming to Harvard. She offered me a Coke and I told her in English that I was soon to give Nasrine away. Not understanding what this phrase meant in English, she replied she understood my decision for people from the Middle East are not our type. After I clarified my remarks, she reversed course and said how pleased she was that Nasrine was marrying into one of Geneva's most illustrious families, one that supplied fragrances worldwide for the perfume industry.

On the evening before the wedding, the bridegroom's father hosted a small dinner for the bridal party at Locke-Ober, Boston's best restaurant on Winter Place. At dessert, he posed for me the most important question of the evening. Was the fair Nasrine a good chemist? Without faltering, I replied that without that attribute Nasrine would not have a place at Harvard. At the wedding reception, held in the dignified Copley Plaza Hotel, messages were read from friends unable to attend. Every-

JDW and Nasrine Chahidzadeh on her wedding day, May 1967

one laughed when one was from a woman I'd met in Geneva. After expressing her wishes for much future happiness to Pierre Yves and Nasrine, she added "let's hope that Jim Watson finds a wife as beautiful as Nasrine." Afterwards, I drove the French-speaking girl with whom Nasrine had shared a flat back to its location on Massachusetts Avenue towards Central Square. There she told me that I was not the only person of note that had visited to collect Nasrine for quiet dinners. Soon after she had arrived from Europe, the junior senator from Massachusetts had phoned her in my lab to renew the conversation he had started some months before in the Firmenich home on Lake Geneva.

But by then I had already found the needed beautiful girl. The blue eyes and full cheeks of the Radcliffe sophomore called Elizabeth Lewis were making me greatly anticipate the several afternoons each week she helped me with secretarial matters. Although she was to go off to Montana for a summer job, I had hopes of her returning to the lab in the fall. Before she left for a few days at her home in Providence, she eagerly accepted my last-minute invitation to a faculty cocktail party that I felt would be dull without her. We drove afterwards in my MG into Boston for a movie and walked together slowly about the Radcliffe Yard. I hesitated to hold her hand then, but realized she might like me more than a little when her postcard came from Montana.

During the fall of 1967 I found Liz even more captivating and introduced her to my father with whom we increasingly had early evening meals at the Hotel Continental basement restaurant. Our first real date was for the Christmas Party in our newly formed Department of Biochemistry and Molecular Biology. In its big basement party room, Paul Doty spotted us and afterwards told me that I had found a peach of a girl. Soon we were holding hands. Wanting to do so no longer in private, I found myself late in March 1968 awaiting Liz's arrival at San Diego Airport. The Harvard spring vacation week was about to happen, and I had been at the Salk Institute for several days at a meeting bringing together cancer researchers and journalists. One reporter chose to interview me, noting that I was nervously holding a can of Coke as I talked about my new position as Director of the Cold Spring Harbor Laboratory. This new job did not require me to leave Harvard but would give me the opportunity to have a second lab where I could start work on tumor viruses. What the *New York Times* reporter did not know was that two days later I was getting married.

Besides Liz and her parents, the only person who knew what I was up to was a highly literate polymath from London, Jacob Bronowski, and his secretary, Silvia. I wanted to get married quickly and simply, and with Silvia's help found just the right cleric, Reverend Forshaw of La Jolla's Congregational Church. He would marry us at an evening ceremony, at 9 p.m. on March 28. Early that afternoon, I took Liz from her plane to a clinic where we were certified free from venereal disease and later we drove 15 miles north for a marriage license. Then we went to the Valencia Hotel, where later we were to spend the night, and had supper at the Whaler's Bar. We then drove to the Bronowskis' strikingly modern oceanside home, where we were photographed many times, and finally to the church to meet the Reverend Forshaw. He suggested a not-very-religious ceremony, and we agreed. I, noting books by Bertrand Russell

With my bride, March 28, 1968

on his bookshelf, suspected he had officiated over many such ceremonies in the past.

Ten minutes later we were married with Jacob my best man. Then we
made the two-minute walk to the Valencia Hotel where we had arranged
at the last minute a reception for my friends at the Salk Institute, among
them Leslie and Alice Orgel who had moved there permanently from
England. Leslie refused to believe I was married, thinking the occasion
was a practical joke and Liz a professional model hired to fool him. The
next morning I sent to Paul Doty a postcard on which I wrote "19 year
old now mine." He had been right in thinking that in Liz at long last I
had the appropriate girl.

Now, more than thirty years later, she remains very much a sweet
peach.

George Gamow Memorabilia

1. See page 32.

July 8th
1953

MICHIGAN UNION
Ann Arbor, Michigan

Dear Drs. Watson & Crick,

I am a physicist, not a biologist, and my interest in biology can be justified, if anything, only by my recently published book "Mr. Tompkins Learns to Facts of Life" (Cambr. Univ. Press. 1953). But I am very much excited by your article in May 30th Nature, and think that this brings ~~the~~ Biology ~~in~~ over into the group of "exact" sciences. I plan to be in England through most of September, and hope to have a chance to talk to you about all that, but

(T.O.P.)

I would like to ask a few[12] questions now. If your point of view is correct, and I am save it is at least in its essentials, each organism will be characterized by a long number written in quadrucal (?) system with figures 1,2,3,4 standing for four different bases (or by several such numbers, one for each chromosomme). It seems ten more logical to assume that different each properties(single genes?) of any particular organism are not determined "located" in a definite spots of chromosomme[*], but are rather determined by different mathematical characteristics of the entire number. (something

---

[*] as assumed in classical genetics

3

Michigan Union
Ann Arbor, Michigan

. . . like the coefficients in
Fourier series). For example
the animal will be a cat if
Adenine is always followed
by cytosine in the DNA chain,
and the characteristics of
a herring is that Guanines
always appear in pairs
along the chains. This would
open a very exciting
possibility of theoretical
research based on ~~mathematics~~
of combinatorix and the theory
of numbers! I am not
clear, though, how such a
point of view would fitt
with genetic experiments,
such as crossovers, which
lead to gene-location along
the entire length of chromosome.

1. See page 32.

But I have of a feeling this can be done. What do you think? Please write to my home adress [Dr. G. GAMOW. 19 THOREAU Drive. BETHESDA. Md.] will be here after July 18ᵗʰ.

Yours truly

G. Gamow.

P.S. If there are only four basic groups attached to DNA - chain, why are the viruses so choicy in selecting there hosts?

P²S. If one puts DNA (extracted from some animal) into the solution containing four bases. Would it reproduce, and, if not, why?

2. See page 58.

San Francisco Overland
Chicago and North Western System
Union Pacific Railroad
Southern Pacific

Feb 7th.

Enroute

Dear Watson,

or (isn't it simpler) Jim,
As you see, I am going
streight to Frisco, and by
train. But I am geting
in Frisco a brand new
whight ~~cheap~~ Mercury @convertable
(will be named Leda) and will
drive over ~~to~~ sometime before
Max leaves for germany
to see both of you. Francis

2. See page 58.

was in wash. a few days after
you, but was too besieged by
DTM people to talk to him much.
   Sitting in my roomette,
looking at Yayoming desert,
and thinking about riddle
of life, I awrrived to a possible
relation between DNA and RNA
which J cannot check however
for the lack of sufficient knowledg
in my head and biol. library
in te train.
   Do J remember correctly
that DNA is completely absent
in plant viruses which have
only RNA ? ☺ J guess J do!
   Question : Is DNA is also
absent in plants in general ?
(J mean : non parasitic plants;
not like bacterias and orchids)

2. See page 58.

J do not know; = but if this is true the following argument
≈ could be made :

steak.

meat proteins are broken up into amino acids and sent to cells.

PARASITE

DNA & RNA

Idaho potato.

Fig 1

vegetable carbohydrites are broken into sugars and sent to cells.

Thus, in parasite the cells must sintesize proteins both from amino-acids and sugars, whereas in plants only sugars are avaliable! *)
Could it be that :

I) amino acids + DNA → contain proteins ?

II) sugars + RNA (+ Nitr. from soil) → other proteins ?

and J don't mean E-Coli who are fed amino acids by Dick Roberts.

2. *See page 58.*

It is most probably wrong
(though, may be, correct in
a way), but I am having
good time anyway finishing
my fourth Scotch & soda,
and looking for a good steak
a bit later (see: illustration
on previous page).

And I had to write
anyway, asking you to
write to me at:

Dept. of Physics
Univ. of Calif.
Berkeley Calif.

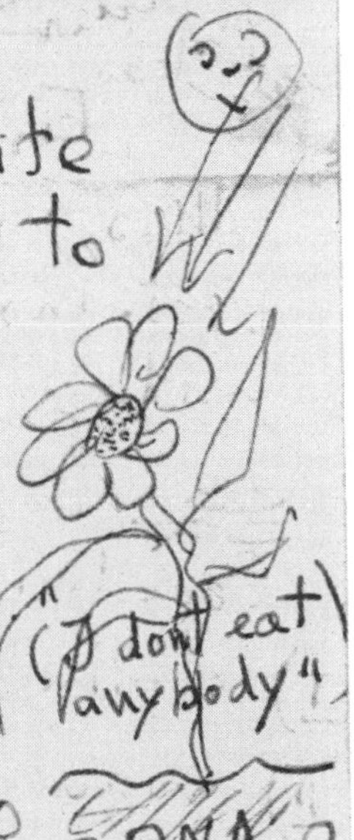

"I don't eat anybody"
? RNA ?
only.

Fig 2.

Yours Geo

(Love to Herr Max).

3. See page 66.

1454

UNIVERSITY OF CALIFORNIA

DEPARTMENT OF PHYSICS
BERKELEY 4, CALIFORNIA

March 7th

Dear Jim,

I have just received the rest of the plastic rings for the bases, and got two boxes of Fisher model balls. Ready to start building DNA, and ordering metalic supports for the sugar phosphate chains. Please let me

know: 1) The exact angle at the axis of the helix between two directions towards the centers of two sugars. 2) The distance between the axis and the center

axis of the helix.

? ?

Sugar   Sugar

of the sugar. 3) The tilt of the plane of sugar ring to the plane normal to the axis. May have the model ready by the time

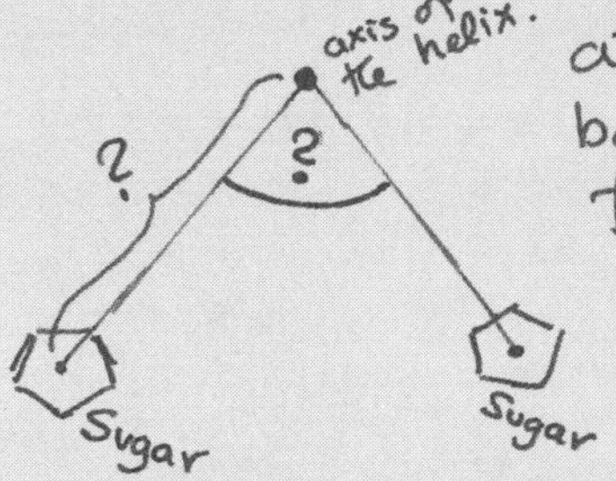

3. See page 66.

you come here.

I am playing now, with 20 triangles (like , Δ_3^2) which may be usefull for RNA. They have rader different combination rules than diamonds. Four of them combine with 10 each. Twelve combine with .7 & each. And four combine only with 5 each. Will tell you more about it when you come here.

     Yours Geo

Please answer quickly.

P.S. The drive along Calif.1. was very beautifull !

4. See page 75.

1954

May 26th.

Dear Jim,

Thanks for your letter. I quite agree that there should be only (20) regular members of ~~RNA~~ RNA - Tie - club. Each member should have a pin (tie-holder) engraved with corresponding amino-acid. You will be probably "Val", me "Glu" ect. There may be some extra members for superfluous amino acids.....

Looking forward to have a lot of discussions on RNA and things on the Muscle Beach. Mass. (incl. ATP- ~~P~~ ~~Ribose~~ -ribosine ect)

Yours Geo.

P.S. I did not hear directly from Aleck or Leslie, but Günter told me today that they may come here this week end. It is too bad because Rho is arriving Friday and this week end (through Tuesday) we will drive Leda to Yosemity (Next week end to Carmel). But, ceterum censeo areneum esse delendam! G.

5. See page 89.

To celebrate my arrival in Woods Hole, you are invited to meet Mr Tompkins and the Facts of Life at a wiskie twistie RNA party in the Szent Gyorgyi cottage on Muscle Beech, Thursday August 12th about 8³⁰ P.M.

Yours

Geo Gamow

RSVP c/o

Professor Albert Szent Gyorgyi

HEADQUARTERS QUARTERMASTER RESEARCH & DEVELOPMENT COMMAND

QUARTERMASTER RESEARCH & DEVELOPMENT CENTER, US ARMY

NATICK, MASSACHUSETTS

QMRDY   8 October 1954

Professor G. Gamow
The Johns Hopkins University
Operations Research Office
6410 Connecticut Avenue
Chevy Chase, Maryland

Dear Dr. Gamow:

    I am very pleased to have received your letter of 23 September 1954 accepting our proposal that these Laboratories help sponsor your biannual conference. We in the Quartermaster are very much interested in the problems which you folks are considering and are most willing to serve in any capacity that may contribute to the success of your meeting. As you may know, the Quartermaster Laboratories have recently been consolidated in new facilities on the banks of Lake Cochituate in Natick, Massachusetts. We feel this will be a most suitable spot for a relaxed and stimulating program.

    Please advise Dr. Ycas or myself as your plans develop so that we may make the necessary arrangements.

                          Sincerely,

                          S. David Bailey

                          S. DAVID BAILEY
                          Chief, Pioneering Research Division

                                Oct 22d

Dear Jim,
    So that is that. In your capacity of the president (par excelance) of RNATiE club, you must now write to Dr. Bailey the exact plans of the spring conference.   (over)

Just got your letter about
RNA production; it looks good
although I don't know chemistry.
What worries me, however, is
that looking through tables
in Chargaffs proof (no chapter
in new volume for Acad. Press), I
find Species for which the
Ad+Th ratio in DNA is high (>1.3
Gu+Cy) the ratio of Ad+Ur in RNA is
Gu+Cy
low, and vice versa. How
would you explain that?

P.S. I would like to have
the TIE by Nov. 8 to wear it
presenting my paper on "Numerology
of Polypeptite chains" (Maniacal
results) on N.Y. meeting
N. Ac. Sc.   Regards also to Leslie
            and Max.
            Yours (Geo.)

7. See page 118.

COSMOS CLUB
WASHINGTON 8, D. C. Nov. 5th

Dear Jim, !

Tie is wanderfull, will wear it at Nat. Ac. Sc. meeting next week, Am sending a chain letter to 17 present (prospective) members of RNATIE club (ordered at random) to decide which of the 20 amino acids they should have on their tie pins *) Hope that the loss of wisdom tooth will not slow down the solution of RNA riddle.

Regards to Leslie, Max, ect

Yours Geo,

*) I will take care of pin production here in Wash. jwelery . . .

8. See page 119.

1954

Nov. 23^d

2121 Mass. Ave.

Dear Jim,

Thanks for your letter. Damned protein sequences still resist, but "ceterum censeo decodeum esse delendam!".

I also put my teeth into ACTH, and got a result (as usually negative). As you remember, my decoding of

‑del̶b̶l̶y̶‑Val‑Glu‑Glu‑Cys‑Cys‑Ala‑Ser‑
‑Val‑Cys‑ (as per Danish Acad. article), lead to a contradiction with the rest of Insulin molecule. However, it turned out that the series is actually ‑Val‑<u>Glu</u>‑<u>Glun</u>‑Cys‑Cys‑eit so that the advantage of two dubble letters was lost, and simple decoding becomes impossible. This could be a hope for diamonds.

But ACTH contains a sequence:
‑Lys‑Lys‑Arg‑Arg‑Pro‑Val‑Lys‑Val‑
and I have shown a few days ago that it is <u>contradictory to diamonds</u>.

8. See page 119.

I have, however, high hopes for[2]
" loose ~~for a~~ triangular code" producing <u>two</u>
"complimentary" proteins like
Ins. A and Ins. B (between which
Martinas & Yčas found a relation).
The scheme ~~code~~ is :

Ins. A.

Ins. B.

What I am trying to do now
is to write a sequence in which
odd members are from Ins A,
and even from Ins B (or vice
versa), and to decode it in
the regular triangular code
which is very restrictive.
But so far, no results.
I have also constructed
the new empirical curve for
the number of different neighbours
vs. accuracy including <u>new</u>
ACTH (altogeder 165 aa), and

8. *See page* 119.

COSMOS CLUB
WASHINGTON 8. D. C.

will have a purely statistical
run on it on Maniac in te
nearest future. Hope it will
show a difference from random.
I am including a sample of
how Maniac did it with old data.

It will be nice to see you here
after New Year. I have to give
a lecture in Univ. of Florida (Gainsville)
on Tuesday night Jan. 4ᵗʰ, and
will be back to Wash. only
on te morning Jan. 6ᵗʰ.
Hope you will ~~can~~ not pass
Wash. before that date. On
te oker hand, I am participating
in a "Rand" conference in Santa
Monica on Jan. 30 & Feb. 1, 2.
Either befor, or after that I
plan to stay for a few day
in Pasadena so that we
can speek more.

8. See page 119.

I am sending your list of participants to Martinas, but there seems to be some finantial difficulties with the Quartermaster. Write to Martinas, and ask about it (adress. U.S.A. Quartermaster. Research Center. Natick. Mass). ← they moved recently.

The included chain letter to the members of RNATIES club is self explanatory. Please send it over to Teller.

Regards to Leslie
Yours Geo

P.S. The situation with Rho, which was deteriating during last two years, became quite bad, and I am living now (temporarily, or, at least, so I hope) in Cosmos Club. Please write to that adress.

P^2S. Spent all Monday in Princeton talking to Bohr and Max about chromosomes.

Please fill it up,
as the original, and
send on. G.G.

[handwritten, date:] Nov. 25ᵗʰ 1954

Department of Physics
The George Washington University
Washington, D.C.

Dear Member of the RNA Club: [SECOND COPY]

I am glad to inform you that the first sample of an RNA tie has been produced by a haberdasher in Los Angeles. They are now available by writing to Jim Watson, Biology Department, California Tech, Pasadena.

The plan we have established is such that each member (up to 20 in number)* can choose an amino acid to inscribe on a tie pin. The amino acids are as follows:

1. Alanine
2. Arginine
3. Aspartic Acid
4. Asparagine
5. Cysteine
6. Glutamic acid
7. Glutamine
8. Glycine
9. Histidine
10. Isoleucine
11. Leucine
12. Lysine
13. Methionine
14. Phenylalanine
15. Proline
16. Serine
17. Threonine
18. Tyrosine
19. Tryptophane
20. Valine

To establish an equitable distribution of amino acids among the members, we have chosen names at random among the 17 existing members, and are sending around a circular letter to let them choose the insignia for their tie pin. This letter will finally be returned to me, and I will prepare the pins.

This letter will be sent chain fashion, to the following, and each member will then put an amino acid beside his name, and cross this amino acid off of the above list. A following duplicate, mailed 1 week out of phase, insures us against plane crashes.

[← The present.]

Leucine   (1) E. Teller, Department of Physics, University of California, Berkeley.

Threonine   (2) L.E. Orgil, Chemistry Department, California Tech, Pasadena.

Tryptophane   (3) M. Delbruck, Biology Department, California Tech, Pasadena.

Glycine   (4) R. Feynman, Physics Department, California Tech, Pasadena.

Alanine   (5) G. Gamov, Department of Physics, George Washington University, Washington, D.C.

Cystine   (6) M. Ycas, Quartermaster Corps Laboratory, Natick, Mass.

*) In case of overflow, six additional members may be added : 1) hydroxyproline, 2) hydroxylysine, 3) Thyroxine, 4) phosphoserine, 5) Tyrosine-O-sulfate, 6) alpha amino adipic acid. But better keep it at 20 !

9. See page 119.

- 2 -

Histidine (7) M. Calvin, Department of Chemistry, University of California, Berkeley.

Serine (8) H.T. Gordon, Department of Enzymology, University of California, Berkeley.

Proline (9) J.D. Watson, Biology Department, California Tech, Pasadena.

(10) A. Rich, National Institute of Mental Health, Bethesda, Md.

(11) F. Crick, Cavendis Lab., Cambridge, England.

(12) E. Chargaff, College of Physicians and Surgeons, Biochemistry Department, New York, New York.

(13) S. Brenner, Department of Physiology, Medical School, University of Witwatersrand, Johannesburg, South Africa.

(14) N. Metropolis, Los Alamos Scientific Lab., Los Alamos, N.M.

(15) Robby Williams, Virus Laboratory, University of California, Berkeley, Calif.

(16) E. MacMillan, Physics Department, University of California, Berkeley, Calif.

(17) Gunther Stent, Virus Laboratory, University of California, Berkeley, California.

I hope you all use air mail so that the process will be completed soon.

Yours sincerely,

Geo. G.

RNATIE - PIN, (project).

G. Gamov

P.S. Please send a post card to me on receiving this note.

(Postcards may be also lost in Mail.)

1954

COSMOS CLUB
WASHINGTON 8, D. C.

Nov. 28th

Dear Jim,
   Sure J would like to speak
in Baltimore in March. The
date of March 18 is preferable
to 17th because in this case
J will not have to miss
my regular univ. lectures on
Thursday. But, if necessary,
this can be done.
   J am glad to hear that DNA
is back in protein building
business, but cannot see at
all **how** you get as much as
256 possibilities. One of the new
AA's should be called WOT-NOT'ic
acid, being abundant in bird's milk.
Bob Ledley has completed the
details of automatic decoding
procedure by means of simbolic
logic equations, and J am
negotiating with Los Alamos
concerning putting it on Maniac.
(Jt may take several days of continous
running!) Regards
         to Leslie. Yours Geo.

11. See page 120.

DNA -ratio in Calf Thymus.

(Table $\overline{VII}$ from tce proofs of chargaffs chapter in E.A.·Nucl. p·563).

$$\frac{Ad + Th}{Gu + Cy}^{*)} = \mathbf{1.30}$$

(18 values varying from 1.22 to 1.41)

RNA -ratio in Calf Thymus.

(table I in Rich and Watson)

$$\frac{Ad + Ur}{Gu + Cy} = \frac{0.162 + 0.157}{0.352 + 0.330}$$

.why **?** why 2.07

*) + M Cy

12. See page 122.

1958

THE JOHNS HOPKINS UNIVERSITY
OPERATIONS RESEARCH OFFICE
6410 CONNECTICUT AVENUE
CHEVY CHASE, MARYLAND

OPERATING UNDER CONTRACT
WITH THE
DEPARTMENT OF THE ARMY

Dec. 6

TELEPHONE
OLIVER 4-4200

Dear Jim,

Here is something new
in decoding. But I am afraid
you will have to ask Dick
Fyneman to explain it
to you!

Do you agree that Ledley
should be elected into
RNATIE club?

Regards
Yours Geo.

13. See page 123.

1954

COSMOS CLUB
WASHINGTON 8, D. C.

Dec. 17ᵗʰ ?

Dear Jim,   Legend for Fig 1.

Tusks:	upper	lower
piram.	right	left
puri.	left	right

?

Thanks for your letter. I do not quite understand your new RNA template model, but it looks to me as a tiger holding a ~~tennis ball~~ rabit in his mgul. But, what is the relation between upper and lower jaws? But, we will talk about it when you are here, and also about statistical results recently obtained by me with Alex and Martinas. Seems to be very little intersymbol correlation when one applic Poisson distribution to Brenner's table. (Fishy story!)

Fig 1.

Next Friday I am leaving for Florida (a combined vacation and Univ. of Flor. lecture trip) and from X-mass day to N.Y. day (both dates incl.) my adress

(over)

will be: Hotel Plaza. St. Augustine. Flor.
Will return to Wash. on the morning
of Jan. 6th but will leave the same
evening for Univ. of Delaware ~~were~~
where I have to spend (and for a
good pay!) Friday the 7th and
Monday the 10th. Since, however,
I have no obligations for weekend, I will come to Wash. from
Sat. lunchtime to Sunday suppertime,
and we will have plenty of time
to talk. On Sat. night. (Jan. 8th)
Dick Roberts*) from D.T.M. (whom you
know) gives a big party for a
mixture of physicists & biologists.
He asked me to be sure to bring
you along. Thus keep that night free.

much better than on the picture!

           Yours Geo.
P.S. Today I have arranged with
a jewelery here for RNA tie pinns.
Looks like:

filled-in gold, (mine) and coast $5 50 + tax.
Takes two weeks to make to order.
I guess I will send a circular
later advising the members of RNATiE
to order their pins directly from
that jeweler. (Schwartz & Son. 1305- F.str. N.W.)
*) He must be also elected in RNATiE - club.

Jim, Don't forget to send me 3 ties.
I will send you the pins for yourself,
Leslie, and Dick, and one extra
for Simmons Geo.

1 February 1955 P.S.

RAND people were excited
about RNA problem!

Dr. L. Blinks
National Science Foundation
Washington, D. C.

Dear Dr. Blinks: How are neighbours?

Several weeks ago I discussed with Dr. Waterman the possibility of
obtaining the support of the National Science Foundation, for holding
a conference on the role of Ribonucleic Acid in the Protein synthesis.
This question is at present of a vital interest for many biologists
and biochemists (and even physicists like myself), and such a meeting
will undoubtedly lead to a useful interchange of ideas in this at present rather confusing and intriguing field.

Dr. Waterman told me that such a conference can probably be organized
in the early summer, and advised me to talk to you on that subject.
The idea of such a conference originated between Dr. J. Watson and
myself while we were discussing RNA riddles in Woodshole last fall,
and on my present short visit to Pasadena we have worked out a program
for such a meeting in more detail.

The conference, which can be provisionally entitled: "The Role of RNA
in Protein Synthesis" should be held in mid-June in the Boston region
which is most convenient because the majority of people involved are
from the Northeastern part of the U.S. The people whom we would like
to invite to the conference are:

Sidney Bernhardt
Navy Medical Research Center
Bethesda, Maryland

Konrad Block
Chemistry Department
Harvard University
Cambridge, Massachusetts

Elkin Blout
Polaroid Corporation
Cambridge, Massachusetts

E. Chalgaff
College of Physicians and
 Surgeons
Columbia University
New York, New York

Paul Doty
Chemistry Department
Harvard University
Cambridge, Massachusetts

G. Gamow
George Washington University
Washington, D. C.

Fritz Lippmann
Massachusetts General Hospital
Boston, Massachusetts

Arthur Kornberg
Department of Bacteriology
Washington University
St. Louis, Missouri

14. See page 132.

Dr. L. Blinks   -2-   1 February 1955

Daniel Mazia
Zoology Department
University of California
Berkeley, California

A. Mirsky
Rochester Institute of Medical
  Research
New York City, New York

Leslie Orgel
Department of Physics
University of Chicago
Chicago, Illinois

Alex Rich
NIH - Bethesda, Maryland

Norman Simons
AEC Project
U.C.L.A.
Los Angeles, California

W. Stanly
Virus Laboratory
University of California
Berkeley, California

Sol Spiegleman
Department of Bacteriology
University of Illinois
Urbana, Illinois

Gunther Stent
Virus Laboratory
University of California
Berkeley, California

J. D. Watson
Biology Division
California Institute of Technology
Pasadena, California

George Webster
Biology Department
California Institute of Technology
Pasadena, California

Martinas Ycas
Quartermaster Research and
  Development Center
Natick, Massachusetts

Paul Zamersnick
Massachusetts General Hospital
Boston, Massachusetts

Assuming that the invited members will be paid regular traveling and living expenses, we estimate that the total cost of the conference will be about $3,500.00.

I will be back in Washington after February 8, and will get in contact with you on this matter.

Yours truly,

G. Gamow

GG:fs

15. See page 147.

P.S. What do you think
about Gales paper in Nature?

May 23
Marine Bio. Lab.
Woodshole.

Dear Jim,
I have arrived here two days ago,
got a small appartement close to the
lab, and am aclimatzing.
When are you arriving here?

——— RNATIE club matters : ———
1) Please send to me, or bring along,
two ties. (for members.
P.S. You beter give me the adress
of that haberdasher in Lons Ang. so that
I can order after you go to Europe.

2) Szent Györgji wants to be
AD, as honorary member. I think
we should do it even though we
planned to give it to Limpman. Do you agree?

3) So for, I distributed 13 RNA pins
(incl. mine). What about Max?
Does he want to have a tie and
pin, or should we kick his out?

Yours Geo.

May 24th 1
Mar. Bio. L
Woodsho

Dear Alek,

I have arrived here on Saturday, and
am settled in my appartement which is
(quite nice one.)
across the pond from the lab. It is nice c
quiet here, and I consider myself bei
in the "rest-house" after Wash. night ma
I find that I would like to have some mor
of my books about which I didn't think
while leaving*? I don't want to ask Rho
to send them since she certainly will fi
it too complicated. Thus, may I ask you
to drop to the house, get these book, and
send them to me which you probably
can easyly do through NIH. They all
stand on a little shelf above the telephone

1) Proteins. $\underline{I}$, $\underline{II}$, $\underline{III}$
2) Nucl. Acids $\underline{I}$, $\underline{II}$ ($\underline{II}$ must have just arrived
3) General Genetics.
4) ~~Paul~~ Pauling's Chemistry.
5) Atomic Nuclei by me and Critchfield.

And throw in Kitchen clock (not for the
discussions, but for cooking eggs) which
I realy forgot.

Will be very gratefull!

Have just finished "Cosmogony" for E.B.
and begining to concentrate on the book,

← but, please, insure !

*)or, rather, thought I will use library copies... (over)

16. See page 147.

Martinas got new data from Knight on N.A
and Prot. consitution of <u>Tomatoe Bushy Stu</u>
which fitts reasonabley well widn his aging
for T.M.V, and TY.*) May be we will entag
that dish, after all

<u>Some dish!</u>

He will probak
come here next weeke
and we'll try to put
best statistics on it.

But Here is truble widn Gale (Nature Apr. 9,
He makes te asignements : **) from
Asp → ACC ; Glu → GUU ; and Leu → AUU. we

	A	G	C	U	Asp. Theor.	Asp. Obs.	Th.	Glu Obs.	Te	Leu
Tobaco :	0·30	0·25	0·18	0·26	3·0	12·0	5·1	9·1	6·3	
Tomato :	0·26	0·29	0·21	0·26	3·4	11·2	5·9	5·5	5·3	
Turnips :	0·23	0·17	0·38	0·22	10·1	4·1	2·3	7·1	3·2	

A very bad fitt indeed !

By te way , if you need to talk to me on
any urgent business call : 1) The Library of M.I
or 2) Falmouth - 1-778-R. this phone is downstair.
from my appartement, but I can hear it ring

Will be in N.Y. on June 11th to qurrel for 30 i
widn Fred Hoyle on transatlantical hook up
(NBC & BBC). Don't miss it, may be amusing

Best regards to Jane

Yours. Geo.

If you *) don't Know dey are:

Ala – 113	Gly – 344	Pro – 334
Asp – 124	Jleu – 133	Ser – 234
Arg – 122	Leu – 123	Thr – 134
Glu – 144	Lys – 233	Val – 114
Glun – 223	Phe – 244	

The remaing five
(least abundant)
are not asigned.

**) Which are, of course, different from those used by Martinas

17. See page 157.

Rnatie Club

"Do or die, or don't try"

OFFICERS

GEO GAMOW · SYNTHESISER
GEORGE WASHINGTON UNIVERSITY
JIM WATSON · OPTIMIST
HARVARD UNIVERSITY
FRANCIS CRICK · PESSIMIST
CAMBRIDGE UNIVERSITY
MARTINAS YCAS · ARCHIVIST
QUARTERMASTER R. & D. LABS.
ALEX RICH · LORD PRIVY SEAL
NAT. INST. MENTAL HEALTH

July 4, 1955

Dear Pro,

This is the first official club circular.
First, the assignments of tie pins (which, as
you know, were randomized):

1) ALA - G. Gamow   8) GLY - R. Feynman   15) PRO - J. Watson

2) ARG - A. Rich   9) HIS - M. Calvin   16) SER - H. Gordon

3) ASP - P. Doty   10) ISO - N. Simons   17) THR - L. Orgel

4) ASN - R. Ledley   11) LEU - E. Teller   18) TRY - M. Delbrück

5) CYS - M. Ycas   12) LYS - E. Chargaff   19) TYR - F. Crick

6) GLU - R. Williams   13) MET - N. Metropolis   20) VAL - S. Brenner

7) GLN - A. Dounce   14) PHE - G. Stent

From this list, 13 members have obtained their tie pins while
the remaining 7 are still stubbornly holding out.

For RNA ties, please write to Jim Watson at Cambridge University.

The first matter of business is the election of honorary base
members. The organization committee proposes two candidates out of
the maximum possible number of four:

    1) Dr. Fritz Lipmann for: CY   (These assignments of
                                        bases were made by
    2) Dr. Albert Szent-Gyorgyi for: AD   random choice).

Each of the 20 members of the club is welcome to send his vote for
both of these two candidates.

For a positive vote: include $1.00 for each candidate (if only $1.00
is included, please specify for which of the two candidates you are voting)

17. See page 157.

-2-

and mail to G. Gamow, M. L. B., Woods Hole, Mass.   For a negative vote: do not do anything.

If the amount of collected dollars is sufficient to buy an RNA tie and a special base pin for the proposed candidates (for, of course, honorary members do not pay) they will be considered electe.

The latest date for receiving the votes is September 1st, 1955.

Sincerely yours, *Ala,*

The Synthesizer

Am leaving for Calif. next week, and untill Aug 20ᵗʰ my adress will be:

CONVAIR. San Diego. Calif.

Since J will be probably dropping to Los Ang. quite often, will you please give me the adress of RNATIE - haberdasher. J may have to order a few more ties.

Yours Geo.

18. See page 186.

1955

Nov 8th
8302 Thoreau Drive
Bethesda. Md.

Dear Jim,

A number of members of Rnatie club are bombarding me with tie demands. Will you please let me know (by return air mail) the adress of this haberdashery in Los Angeles where they are being made.

What do you think about this Rundle's paper? Should we elect him as an honorable member of the club, or should we disband the club altogether?

I think I will go back to Cosmology! I think that most interesting thing about this RNA model is that it explains that all proteins (Oxitocin, Vassopresin, Ins. A, Ins B, ACTH, Ribonucleas, and the recent "melanophore expanding" hormone *) have $3n$ amino acids. This is almost exactly Barbara Low's hypotesis of triple reding of non punctuated template. Oh gosh!

9, 9, 21, 30, 39, 126 & 30.

* also 135 in TMV! Love to Francis. Yours Geo.

19. See page 187.

RNATic Club
"Do or die, or don't try"

Nov. 15th 195.
8302 Thoreau Dri
Bethesda 14 Ma

Dear Alex,

Thanks for your letter, and incl....
Congratulations on the collagenes and
love to Francis... I had a very nice
summer in Woodshole, wr wrote more
than half of the book (11 chapters out of 21)
and two nice (addapted) limerixes:

King Dan–el Mazia of Sheba   Queen Gertude Mazia of She
Was in love with a tiny amöba.   Was jolouse of tiny amö
This whee bit of jelly   Said she: "Ain't it od
Had crowled o'ver his belly,   That this damned pseudge
And metabo whisper "Ich liebe".   Entführte der Man dem ich l

Upon my return to Washington at the en
of September, stayed for a while in Cosmos
Club, but moved into the house (as above)
three weeks ago when Rho moved out int
a downtown hotel. Am busy now stripping
the house down for selling, packing books
and things worth storing, and going
through all this dirty business. The thing

(Pto)

19. See page 187.

L

Rnatic Club
"Do or die, or don't try"

are geting horribly complicat and utterly unpleasant, for one thingk because my attorne who is organizing settlement and divorce has to coordinate all his moves with Rho's psychatrist.... God Knows how and when it ends! I am back to Wash. only seven weeks and am dreaming of getting out of here as soon as possible, and as far as possi. One of the possibilities is to get an (advanced) sabathical from G.W. and go to England at the end of this semester(and through the summer. I may coordin it with ORO which has an office in Londo. But, it all will depend on legal angle of maintaining the residence in Maryland for all this damned business...

As you see, I am not in very good mood, or state, and the only things I can still do is to go on with my lecture (dropped one of them tonight).

Yours as everg Geo

Love to Jane

P.S. What do you think about this midwestern HNA model?

20. See page 190.

Rnatie Club
"Do or die, or don't try"

Dec. 1st 1955

COPY.
(J fight for you, boys! Geo.)

Dear Dr. Rhoads,

Thank you for your kind letter of Nov. 23d. J think, however, that it was based on some misunderstanding. Jn my article in Scientific Amer. (Oct. 1955) J do not, as you incorrectly quote: " give credit of the studies concerning the molecular configuration of DNA entirely to watson and Crick ". Jn fact, the text runs : "The first question (i.e. How the information is stored in chromosomes ?) has already been at least partially answered thanks to the work of a number of investigators, notably the team of Crick and Watson". And, in fact, the present model of DNA (structure) and the hypotcesis concerning its selfduplication was first proposed by W &C. as it is also stated in the begining of Nature's

(by 9 authors)
article of May 1955 which I have, of course, read when it appeared last spring. In fact, it says: " W & C have proposed a structure .... and from this structure derived a hypothesis concerning the selfduplication of nucleic acid." Of course, in a popular article in Scient. Amer. I could not give a complete list of all publications on that subject, and, in particular, could not refer to the institution which, as you say, "has provided much of DNA" used by Wilkins and Associates in their X-ray studies. But I do give the reference to Crick's article in Oct 1954 Scien. Amer. in which he writes: " Watson and I were convinced that we could get somewhere near DNA structure by building scale models <u>based on</u> <u>the X-ray pattern obtained by Wilkins,</u> <u>Franklin and their co-workers</u> "

Thus I think that all properties are
are ~~retained~~ preserved, and no offence to anybody ~~is made~~ ment.
with best personal regards
Yours truly G. Gamow

21. See page 239.

Rnatie Club

"Do or die, or don't try"

Nov. 24^d 1956
Boulder. Color.

Dear Jim,

Well, after long gipsying I am finally settled here, and am aproaching asimptotically the state of equilibrium ..... (it is wonderfull here !)
I have at the moment no new ideas on RNA decodiny, but .....

In the middle of August I am planing to organize here a conference (sponsored by NSF) on "Molecular Genetics".

COULD YOU COME?

Ted Puck doing wanderfull things with his mamalian celles. A recent result is that the celles from different tissues replicate as a true mutations. Thus, basic changes must have taken place at the stage of bastula, gastrula, or ...

How are you Jim?

Yours Geo.

22. See page 239.

UNIVERSITY OF COLORADO

BOULDER, COLORADO

DEPARTMENT OF PHYSICS

Dec. 6ᵗʰ 1956

Dear Jim,

Thanks for your letter. The conference will be definitely during the second week of Aug. since it is the only time when Muller can be here. I will not be able to invite all RNATIE members since many other must be invited and the fonds provided by NSF are limited. When Ted Puck and I will have a list ready, and coordinate with NSF I will send you a copy.

I am fairly busy giving last touches to my book "Matter Earth and Stars" which I started in Woodshole, and have not thought much on biological topics recently. I have collected £1000 in U.N. in N.Y. recently, but will go to India for a couple of months only next winter. But I am going to spend these X-mas vacations in Caracas delivering a few lectures (in spanish!). I will be in Cambridge on March 4ᵗʰ and 5ᵗʰ and hope to see you at that time.

Yours Geo.

ILLUSTRATION CREDITS

frontispiece, pages 21, 26, 33, 68, 112, 230, 238, 254, 256, 258:
   Cold Spring Harbor Laboratory Archives
page 9: Cavendish Laboratory Archives
page 17: Nature Publishing Group
page 22: Seymour Benzer; Cold Spring Harbor Laboratory Archives
page 24: Norton Zinder; Cold Spring Harbor Laboratory Archives
pages 27, 184, 191: courtesy of Christa Menzel
page 38: courtesy of Derek Burke
pages 41, 43: California Institute of Technology Archives
pages 57, 155, 179: courtesy of Alex Rich
pages 63, 69: courtesy of Igor Gamow
page 81: National Library of Medicine, Bethesda: Institute for the History of Medicine
page 99: John Engstead; *Vogue* magazine, Condé Nast Publications
page 111: courtesy of Emma Rothschild
page 134: courtesy of Erwin Chargaff
pages 140–141: courtesy of Francis Crick
pages 174: Eleanor and Clyde Moore, 2001 (www.photosbyeleanor.com)
pages 175, 237: West of Scotland Press Agency
pages 205, 224: courtesy of Don Caspar
page 229: Francis di Gennaro; University of Maryland at Baltimore
page 244: Melanie Jackson Agency—Feynman Estate
page 245: Rick Stafford; Harvard University News Office

A NOTE ABOUT THE AUTHOR

Born in Chicago, James D. Watson studied at the University of Chicago (B.S.) and Indiana University (Ph.D.) before going to Europe in fall 1951. After a year in Denmark, he moved to the Cavendish Laboratory in Cambridge, England, where he met Francis Crick. For their 1953 discovery of the double helix they, with Maurice Wilkins, were awarded the 1962 Nobel Prize in Physiology or Medicine. In 1956, he became attached to the Biological Laboratories of Harvard University, where he remained a member of the faculty until 1976. In 1968 he also began serving as director of the Cold Spring Harbor Laboratory on Long Island, New York, and shifted its research focus to the study of cancer. Since 1994, he has been its president. Between 1988 and 1992, he was associated with the National Institutes of Health (NIH), helping to establish the Human Genome Project. His seminal textbook *The Molecular Biology of the Gene* was published in 1965. He later helped to write *The Molecular Biology of the Cell* (1983) and *Recombinant DNA: A Short Course* (1983). He is also the author of a previous memoir, *The Double Helix*.

A NOTE ON THE TYPE

This book was set in Fairfield, the first typeface from the hand of the distinguished American artist and engraver Rudolph Ruzicka (1883–1978). In its structure Fairfield displays the sober and sane qualities of the master craftsman whose talent has long been dedicated to clarity. It is this trait that accounts for the trim grace and vigor, the spirited design and sensitive balance, of this original typeface.

Rudolph Ruzicka was born in Bohemia and came to America in 1894. He set up his own shop, devoted to wood engraving and printing, in New York in 1913 after a varied career working as a wood engraver, in photoengraving and banknote printing plants, and as an art director and freelance artist. He designed and illustrated many books, and was the creator of a considerable list of individual prints—wood engravings, line engravings on copper, and aquatints.

Composed by North Market Street Graphics, Lancaster, Pennsylvania
Printed and bound by Quebecor World, Fairfield, Pennsylvania
Designed by Robert C. Olsson